Veteran Wealth Secrets

The Post-Military Playbook for Obtaining Autonomy and Financial Control in the Modern Economy

Scott R. Tucker

ISBN: 978-0-578-81320-2

US VetWealth

Email: support@usvetwealth.com

Dedicated to Mom, Jen, and Stephanie, the only true believers since day one.

Also By Scott R. Tucker

The Veteran's Guide to Life, Liberty, and Purpose

The Modern Guide to Veteran and Military Life Insurance Planning

The Veteran Retirement Rescue

Looking for Increase?

Get all of the *free* resources mentioned in this book at:
usvetwealth.com/resources

Subscribe to *Veteran Wealth Secrets* on YouTube

Apply to be a guest on our podcast at:
usvetwealth.com/guest

To inquire more about our services, schedule a 15-minute discovery call at:
usvetwealth.com/discovery

If you know you have something unique and valuable to contribute to the world...

If you feel like doing the "right" things you've been told to do isn't the right thing for *you*...

If you believe that life is not about money...

If you want to stop trading your time for money...

Then this book is for you.

I didn't write this book to try to convince you. I wrote this book to share my message with those who already know that what I write about is true. If it happens to convince you, you're welcome :)

Table of Contents

Preface

This is not another military transition book. That's already been done. Other authors have written a lot of great books that are very specific to the typical military transition and the tricks to get ahead.

This is not a traditional investment book about military millionaires or Thrift Savings Plan (TSP) millionaires, nor is it the veteran's guide to financial freedom. Those books are very specific to just saving more money. While helpful for many, they're not really strategic beyond budgeting, saving, and investing.

This book is probably closer to the book *Rich Dad, Poor Dad* by Vietnam veteran Robert Kiyosaki. He talks about self-employment, business ownership, and investing in income-producing properties. For example, real estate investment is sexy to a lot of military and for obvious reasons. If you're smart and you're stateside, you can use your VA loan to buy your first house and then rent it out after you PCS. I don't always advocate most active duty military using the VA loan to buy a house because they're not going to do what's advocated in *Rich Dad, Poor Dad,* or they're just not going to like it. I didn't like owning property. Personally, I'm just not a real estate guy. Instead, I'm going to teach you to invest in yourself. Most people think that means going back to college. You can spend years of your life and hundreds of thousand dollars for an education—unless you use the GI Bill, of course, in which case it's free. Unfortunately, people don't take free things very seriously, and far too many veterans end up using their GI Bill and don't really get any return on the taxpayers' investment.

But I'm not talking about college. I am talking about investing in the human capital that you already inherently have. You, as an individual, have value that you can bring to others independently of any college degree or certification. Tapping into that human capital requires a little bit of creativity. It requires a little bit of thinking outside the box. It also requires some financial preparation. But when it actually comes to making money, it's a far better investment strategy than doing stocks and bonds and investing in traditional retirement accounts. It's actually even better than real estate investment. I think real estate investments are great opportunities for some. Starting a business is an awesome opportunity for others, so is just being self-employed in a service or affiliate capacity for someone else. But this strategy that I'm going to teach you is even better than starting a business, but only because if you just do this first, it will make all the other things easier.

Of course, as you're reading this book, you may think, I can't do this. It's not for me. Some of it may not be for you. But if any part of it makes sense for you, if you get one epiphany, and you want to implement it, I'm happy to help you. I wrote this book as a way to introduce my philosophies and strategies to potential clients. In it, I show you exactly what I'm doing and how you can do it too. If you find that provides value for you, and you want to do these things too, great! That's how I make a living. If you decide my approach is not for you, then I hope you still gained something from reading this, and I'll be happy that neither of us wasted time trying to conduct business that would not have been successful.

As you read, you're probably going to be thinking to yourself: if your (financial and non-financial) investment strategies are so great, why isn't everybody doing them? Well, why isn't everybody doing real estate investing or buying into a franchise, for example? My strategies are meant for those who really want to take control of their human capital and leverage not only their military benefits, but the other financial vehicles that are available to all Americans based on the U.S. tax code and the financial regulators. My strategies are meant for those who want to truly take advantage of growth without risking their money or losing it all to taxes.

My friend and Marine veteran Jason Stapleton preaches over and over on his podcast *Wealth, Power and Influence* that when it comes to investing, there are only two rules:

1. Don't lose money.
2. Invest in things you understand.

I believe this to be true. Unfortunately, 99% of what most financial advisors will steer you into comes with a huge risk of losing money. Even worse, 99% of investment advisors will keep that knowledge from you—along with the fact that they don't truly understand what's going on in the stock market themselves. They are required by law, as am I, to disclose on any and all marketing material that "past investment performance does not guarantee results." Translation: "It's impossible to predict the future, Mr. Investor, so the risk is 100% yours and it'll cost you 1.5% in asset management fees regardless of how your investment performs. Oh and by the way, if this is a taxable account, we're just gonna pass those taxes on capital gains off to you. Thanks!" I know that this is the case because I was in the industry for 10 years. It took me 10 years to really see it. It also took me 10 years to realize that the people who have been in the financial industry for 30 or 40 years don't even give a shit that they're essentially gambling with their clients' money. They're okay with the status quo: "Ehhhhh, sometimes clients make money, sometimes clients lose money. Either way, we're going to make our products and services sound as good as possible in our marketing."

Really, though, it's all a crapshoot.

But the one thing that you can understand is yourself. We're not taught that in school. We're not taught that in society. Instead, we're taught to conform, and conformity breeds apathy. For the less than 1% of financial advisors who actually get this, we know that we are doing the best thing for our clients. We know that we are being a true fiduciary because we're doing what's in our clients' best interests and helping them to create the wealth that they deserve. The other problem that is rampant in the industry is that most financial advisors assume that wealth is about ones and zeros on a bank account statement when really, money is just a means to an end.

What I'm going to teach you in this book is an opportunity to take advantage of the income you're making now, especially if you're still on active duty. On the one hand, I'm going to show you how to sock money away where it can grow without risk and where you will still have the ability to access it whenever you want and without having to waste a bunch of time trying to manage the taxes (because this strategy isn't reportable to the IRS). Why? Because you're going to need money for the biggest opportunity of your lifetime: the moment you leave the military.

The moment you leave the military is the moment that you regain the liberty that you spent so many years serving to defend. I mean, let's be honest: when you're on active duty, you don't have any freedom or liberty. You get paid well, believe it or not, and if you do what the military millionaires and the real estate investors say you're gonna be way better off than the average veteran. But if you do this first, it's going to make it a lot easier, a lot cheaper, and a lot more accessible for you to experiment with these various opportunities so that you aren't just stuck in a fixed-income job that you end up hating and eventually resent.

You know, one of the strangest things I've noticed over the years is that while so many who are serving a career in the military and want to get their pension must love serving, we all know that the time we get doing jobs in the military that we truly love are only a fraction of our military career. Most of the time, we're sucking it. Yet why, then, do so many say, "Oh, I'm gonna stay for another three years and see if maybe I get promoted and can get a little bit of extra income on my pension," when the opportunity cost of those three years is completely overlooked?

The opportunities we have today, in the Internet age, are incredible. This age of opportunity started around the turn of the century, and the opportunities have only grown after 20 years of innovators and early adopters who have just made it so damn easy to make money. The amount of income that you could create for yourself in three years would blow that slight increase in your pension away. But the reason most don't see the possibility and don't do it is because they haven't done this strategy I'm going to share with you. They keep listening to the veterans who went before them about how they learned to game the benefits system to get a little more from the taxpayers instead of focusing on creating value for others. They haven't positioned themselves to see how to create multiple streams of income and how to leverage their own human capital instead of more benefits, bank loans, mortgages, and venture capital. They haven't positioned themselves to create and then take advantage of these modern opportunities.

As I'm writing this, we're in the second stage of an aggressive lockdown because of the coronavirus. No matter when you are reading this, it has already become clear that this is much like a 9/11 event for our society: at some point, the virus will eventually no longer be an imminent public health threat, but it will leave behind lasting changes in our economy. Essentially, the whole country is in transition right now while we figure out what those lasting changes are going to be and how our families and businesses are going to navigate them. I fear the most for those within the veteran community who aren't prepared to transition again.

There's an underlying complacency when we're in the military. We know we aren't going to get fired, and we can rely on that paycheck. But there's *huge* risk around a lot of people's paychecks now, and I know many veterans have already been affected by it. Part of this book is to teach you how to position yourself mentally for the inevitable, how to position yourself to focus on your strengths, how to develop yourself and your identity so that you can capture opportunities to build and use wealth, which really means opportunities to stop selling your time for money and start spending your money on time to do whatever you want. This book is meant to teach you the financial strategy that will give you that access without slowing you down with banks and taxes and risky stock markets.

In Chapter 20, I tell the story of Commander Jones who, in less than a year, saw himself truly transform into somebody who could create opportunity for himself. (If you don't have time to read this whole book at once, I suggest you start there.) He's not the only one, of course. Thousands of people, every day in America, transform their lives. There are over 1,700 new money millionaires every day! These aren't people who inherited money from their parents. These are people who are creating wealth for themselves. I look forward to seeing who you become at the end of this book.

Unfortunately, most veterans are told to stay in the mindset that they're only meant to serve someone else, they're not meant to be wealthy, when ironically, it's wealth that allows you to serve many. That's how I felt for so long. Once I became self-employed, whenever anybody asked me my income goal, I usually told them that if I could earn what I had been earning in the military, that would be good enough. What a shame that so many veterans think like that when there's so much wealth and money just floating around everywhere—trillions of dollars. It's impossible to fathom how large that number is, yet the reality is that if you manage to get your hand on $10 million of it, no one would even notice. $100 million? It's a rounding error. Yet we are led to believe that making a six-figure salary is some sort of a milestone, and most stop there. Have you ever aimed big for something, fallen short, and realized that you still ended up far better off because you aimed high? Why don't we do that for ourselves when it comes to how we want to live our lives after the military? Worse, our government is wasting so much money. We know all too well how much fraud, waste, and abuse is going on as they continue to send us off to fight endless wars.

Why, when we get so much in pay and benefits, do we allow the government to take some of that from us? Do you even realize that's happening? Why do we just trust the Wall Street marketing machine with our money, and let the financial planning industry take so much from us in fees, when they actually have no idea what's going on? It's because these are the tools we have to play with. This is the game we are in. So let's be real. Let's figure out how to get that game to work to our advantage.

If you go to Vegas, and you play the slots, you've got very little chance of winning. If you play craps, sometimes you might get on a roll and it seems like things are nice—until you roll a 7. Those are games of straight chance. But if you're playing poker, you're not playing against the house. You're playing against the other players, right? That's what's happening all around us. The vast majority of us are playing slots. At least get enough knowledge about what's going on around you so that you can participate and play some poker. Then, if you lose, you'll know why you lost, and you'll learn what to do better next time. We do that in the military. When we have a battle that doesn't go so well, we do an after action review and figure out what we could do better next time. This isn't my first attempt at writing this book. It's my third!

What a shame that our military financial education doesn't do the same damn thing.

I have frequently told the story of something I witnessed at an invite-only Digital Military Experience Conference. The head CFP for a major financial institution showed a slide and said that for the seventh year in a row, over a third of their members, most of them active duty or veterans, reported living paycheck to paycheck. That's a third of about 13 million people! The whole room full of the top military financial bloggers and influencers said, "Oohhhh, Ahhhh, what a shame," and then the presenter just moved on to the next slide. WTF?

Nobody had a solution. Nobody even realized that this problem is already being solved. Part of the problem is that typical financial education is not motivating. The Financial Independence, Retire Early (FIRE) movement is interesting, I guess. Maybe it's great if you can retire at 30, but for what? Just to surf and ride around in an RV?

Navy Veteran and author Doug Nordman is an inspiring example of a better way to do this. He fully retired at 41. He gives away all his knowledge for free. He doesn't just surf, he shares a tremendous amount of financial education. He helps people understand all the nuances of the military benefits. When I first realized I wanted to be this kind of person, that that was the kind of life I wanted to have, I thought that I didn't have enough to offer, that I didn't know enough. But over time, as I learned more and more about my field, I realized that where I really fit into this picture was that I could fill the motivation gap. I could motivate people to see the opportunity to take advantage of something different. My strategy requires a different approach than what folks like Doug teach. I'm not against their approach. That's for some people. My approach is for others. If you are one of the others, then this book is for you. I honestly believe that this is only for about one to 10% of our population, but don't they deserve to know about it?

Of course they do.

The people with whom I think my strategies will resonate the most are the type of people who want to be entrepreneurs or self employed at some point. People don't realize that so many entrepreneurs fail because they didn't set themselves up properly for it. You can't do it overnight. It takes years. And the more years you can prepare for it while you're still on active duty, or while you're getting an income, or while you have assets that you can use to build that foundation for yourself, the better your chances of success are going to be. You won't get stuck trying to get more degrees or more certifications. You're just going to focus on yourself and your own human capital and your financial resources. Veterans have amazing assets and benefits that you can use as a seed to grow instead of just spending it, and being the person saying, "I'm sorry, I'm in a bind right now, I can't do anything, I can't afford anything, blah blah blah."

If you're always trying to save a nickel, you're never going to make a buck. This book is about leveraging the free market to grow and capitalize on the benefits that you earned as a veteran in the post 9/11 era. It's about making a buck and taking financial control back for yourself.

Veteran Wealth Secrets

government hypnotizes you with once you drop your papers are all presented from the perspective of money. The resume-writing guidance is about money: gotta get a job, gotta replace that paycheck! This book is *not* about money. This book is about identity and transformation. This book is about getting to a clear-eyed understanding of who you have been, what you want, and who you want to become. Those are the things that you have to get right, and if you do, money will come along with it, and there will be plenty to meet your needs.

I will get into some of the details of my own personal journey that led me to where I am today a little later. But before I do, in the first part of this book, I want to talk to you about apathy.

Are you apathetic? Do you even know if you are? Most people think of apathy as a state of listlessness and an absence of action, and people who get up every morning and meet their responsibilities—people who go to work, pay their bills, provide for their families, and do the kinds of things that citizens in this country are expected to do—would not consider themselves apathetic. But apathy doesn't mean that. Apathy means a lack of interest and enthusiasm. Do you jump out of bed every morning eager for the day ahead and excited to do work that is meaningful to you and that you know adds value to the world? If the answer is no, then you're apathetic. I'm going to talk about how so many of our country's best, brightest, and bravest got to the point of apathy in the first place, and what that apathy is costing us as a nation and you as an individual.

Then, in the second part of the book, I'm going to tell you how to get out of it.

The second part of the book is about how to get out of apathy and into a state of aspiration: your aspiration. What identity do you aspire to? Who do you want to become? Who are you? I'm going to talk to you about what steps to take and what that transformation looks like in a person's daily life.

to get there. Others have a pathological fear of suits, commutes, or cubicles. Whatever has brought you here, welcome to a supportive community of like minds.

As I write this book, our country is in the midst of the coronavirus pandemic. Schools are closed, as are many "non-essential" businesses. Fifty-five percent of restaurants have closed permanently. The rule of the day is "social distancing." Some aspects of the economy have come to a screeching halt, while others are booming. People are afraid of getting sick. But underneath that fear is another, much more ominous one, because the reality is that for most people, even if they do get sick, they're going to get better. The question is, is the life they have built for themselves still going to be intact when all of this is over? For some, unfortunately, the answer is maybe not. The pandemic is bad. What follows it, in the long run, may be much, much worse.

And yet, my life has not changed one little bit. I have gotten up every morning since the shit hit the fan and done exactly what I was doing before coronavirus became a household word. I have not lost any money, any ability to work, or any sleep.

Why? Because my life is built on a foundation of wealth and liberty, so I am insulated from the social and economic crisis that is gripping our country. Want to know how I got here? This book is going to tell you. But before we dive into it, I ask that you spend a few minutes to think about a few things. You don't have to have all the answers right now, but these are things I want you to keep in mind as you read:

- What dream have you been putting on hold while you served your country?
- What value do you now want to add to the world?
- What passions can money help you to enjoy?
- What do you want your life to look like beyond the military?

Typical military retirement planning revolves around money. Those PowerPoint briefs about your "benefits choices" that the of your life following their orders, all the while socking away as much money as you can into a retirement fund that you can't touch until you're in your sixties. If you do all of these things in exactly this way, you will be doing the smart thing. You will be successful, and you will be safe.

This myth has nothing to say, and no interest in, whether or not you will be happy. Those who perpetuate this myth have no interest in your happiness, your sense of fulfillment, or your own unique value proposition. They are only interested in what you can do for them, and the best thing you can do for them is to cheerfully take your place as the next cog in their wheel and stay there until it's time to retire. Because at that point, you're of no use to them anymore.

Sound grim? That's because it is.

The thought behind the original title was that this book would reveal this myth for what it is and show my fellow service-members and veterans a better way to move into the next phase of their lives, a way that leads not to mediocrity, but to wealth and liberty. And that is what this book does; but when I really got to work on writing it, I realized that I actually didn't want to focus all that much on the myth part of the situation; I wanted to write about what life looks like when it is not constrained by myth or expectation. I wanted to write about clear vision, informed action, and the existence and power of choice and free will. I wanted to write about the real meanings of wealth and what real security looks like. I wanted to write a manifesto.

This book is that manifesto.

If you're reading this book, something has brought you here, and I'm guessing that it isn't something good. People who tend to find me are coming from places of doubt, fear, confusion, frustration, or some combination of them. Some just have a nagging sense that there has to be something else for them, there has to be a better way to return to civilian life. Some know exactly what and who they want to be, but they don't know how

Introduction

Our job in this life is not to shape ourselves into some ideal we imagine we ought to be, but to find out who we already are and become it.
~ Steven Pressfield

The original name of this book was going to be *The Post-Military Retirement Myth*. Throughout history, human societies have been guided by their myths, and our modern world is no different. A myth is a traditional story that explains something about the way the world works. It is a story that is told over and over and over again, and people shape their lives according to its supposed wisdom. Many military service-members and veterans shape their lives according to a whopping myth that's consistently fed to us by the government representatives who "help" us prepare for our transitions back to civilian life. Characteristic of a myth, their story sounds reasonable, it soothes our fear of the unknown (just do this, in exactly this way, and everything will be okay), and it is all-pervasive—everybody believes it. Don't know what myth I am talking about? I'm talking about the 1980s approach to getting out of the military, and it goes something like this: Translate your military experience into the skillset language that the civilian world will recognize. Write your resume. Put on a suit and go to job fairs. Get hired by a government contractor or major corporation and spend the next twenty or twenty-five years

Finally, in the last part of the book, I am going to talk about where achieving that aspiration is going to lead you: to autonomy. Autonomy means "the right or condition of self-government." As military service-members and veterans, autonomy means the right or condition of *not* following orders about what you are going to do with your life, so you can start giving them instead. A state of autonomy has three distinct pillars, and like anything that is meant to stand on three legs, all three of those pillars must be in place and in a state of equilibrium with one another in order to create a solid and secure foundation. I am going to share with you what those pillars are, both in philosophical and practical terms, and tell you how to create them.

Autonomy means that no matter what happens in the world around you, you cannot be knocked off course. The great thing about autonomy is that you get to define it on your own terms. You get to decide how it looks and what it means to you. Your version of autonomy will be as unique as you are. You can be the person you were meant to be and live the life you were meant to live. You can achieve your dreams.

If you picked up this book, it's because you perceive on some level that you have a problem to solve. Since you can't solve a problem without a clear understanding of exactly what the problem is, the first thing we are going to do is define "the problem." Then we're going to get into the really fun stuff.

Ready?

Then let's dive in.

Part I

Out of Apathy

1

The World as We Know It

Blind belief in authority is the greatest enemy of truth.
~ Albert Einstein

I am a veteran. I have many friends and acquaintances who are veterans. I have worked in the financial space since 2008, for clients who are almost exclusively service-members and veterans, and I have created a business that specifically serves the same. It is no exaggeration to say that I have talked to thousands of veterans about their transition, and what I have found is that as we're leaving the military, the vast majority of us are okay with the status quo. In fact, we don't even know there is anything else out there to pursue. Sure, we know that people start businesses, and that sounds interesting, but most veterans have become institutionalized. They are so used to following orders that they have lost touch with their capacity to question what is going on around them.

It's not hard to understand this. The first job of the military is to unmake us. The military dismantles our personalities and rebuilds them to suit their own purposes. They have to—not just for the success of the mission but to give us the best chance of staying alive. We enter the military as individuals, and we leave it as service-members who are so well-trained to follow orders that we have forgotten there is any other way to be. We may leave the military, but our minds remain institutionalized. How do I know that this is the case? Because we follow orders (retirement and transition checklists and trainings) about how to get out of the military, and we follow orders about how to set up our civilian lives, even though there is no necessity for doing so. We have no idea that if we choose, we can see the world through a different lens. Your worldview can and must change if you want to create autonomy. You'll find that this change can be both enlightening and frustrating at the same time.

You may reflexively say that you're not one of the people I'm talking about, because no one thinks of themselves as being "part of the status quo." I mean, that's another way of saying that we're average, mediocre. And of course we're not that. Are we?

We'll get to that a little later, but first, we're going to have a little history lesson because you can't really understand where you are and where you're going if you don't know where you've been. Apathy is not a natural human condition. It's a condition that is created to serve a purpose ? someone else's purpose, that is. Let's take a moment to look back at how we got to where we are today.

How We Got Here

From a historical perspective, we need to realize we are all part of what has been called "the longer now," meaning we're not just born today and then live the life we live. How we live our lives is based upon how those who came before us lived their lives, and the way people will live their lives in the future depends on the choices that we make today. Looking at our lives through the lens of the longer now helps us to answer questions like, "What do we know about how life is 'supposed' to work?" and "Why are things the way they are?"

During the Industrial Revolution back in the early 1800s, when technology created standard jobs and big companies, the new corporations wanted to keep workers happy, so they designed the pension system. Then, during the Civil War, when they needed men to fight, they offered the same thing: come fight for a wage and then get a pension after your service. Every time our country entered a major war, the government expanded these programs. Around sixteen million Americans served in World War II, what was then approximately 10% of the U.S. population. The government created the GI Bill to educate all of these young veterans, and our modern economy evolved from the post-war lifestyle that these benefits created. The path was clear: Get a degree, get married, buy a house in the suburbs. Dress up nice and go to work. We've all seen Mad Men and Leave It to Beaver. Many people believed that this lifestyle was the way to be happy. We built the largest economy in the history of the world. We kept everyone distracted with sports and Hollywood. We didn't pay attention to the principles of liberty, and we allowed Wall Street and the politicians to hoodwink us into ratcheting up the debt more and more. We still have the benefits system that formed the backbone of this way of life. In fact, it has expanded exponentially. Today, we think of benefits as a necessity. We can't imagine life without them. Many of us won't consider taking a job that doesn't offer them.

And where has all of this gotten us? In recent decades we have seen more than one absolute crisis, but no one has tried to solve the problem by addressing the root cause. Our leaders just keep covering the problems up and kicking the can down the road while American citizens look the other way. We've been at (undeclared) war for almost twenty years now. Military service has become just another job. Yes, there's the element of danger, but these days, service-members are not that different from firefighters and the police. We have become on-call workers who get summoned when needed to go do someone else's dirty work. Most of this country is just going through the motions every day in a state of apathy, which includes not seeing and recognizing what is really going on. Moving into a state of autonomy requires that we see and recognize where we are today and what it means for us and for the future. But we'll get to that.

If this is where we have come from, then how else is a soldier—who's used to following orders, standard operating procedures (SOP), and rules and regulations—supposed to think about post-military life? It's nobody's fault. In the post-WWII years, it made complete sense to go get a good job, raise your family, get a house in the suburbs, and be a good citizen and a good taxpayer. So if we think about it, veterans are following these post-war people, the Greatest Generation, as we should. They sacrificed a lot for us. They sacrificed not just in war, they sacrificed their individuality, which gave us the infrastructure to create so many greater and better things.

But turning military service into a job with benefits also turned out to be something of a double-edged sword. On the one hand, war brings massive advances in technology—all the stuff we enjoy today. But do you think that generation wanted their kids and grandkids to end up fighting perpetual wars? We willingly sign up to serve, and maybe in the back of our minds we want to see if we can handle it. But how many times have you been deployed in the past two decades? Is this what the Greatest Generation wanted for us, to be involved in non-stop overseas intervention at the cost of trillions of dollars? No one has said, "Hey, this is nuts. I'm not going to fight your wars anymore." Why? Because we get a pretty damn good paycheck and benefits for going to war, that's why! We will touch on these benefits as they exist today and how they, too, are perpetuating apathy a little further on.

In the past decade, we've seen another monumental change. Decades ago, companies in the private sector got rid of pensions and went to the 401(k) retirement-planning model. It wasn't until recently, in 2018, that the military, too, began reducing the pension. They couldn't reduce it totally because they'd freak everybody out, but they can't afford it anymore. The military has started to act like a corporation.

Service-members are already accustomed to following orders and depending on a paycheck and benefits, which makes them excellent candidates for the corporate world, especially the ever-growing government contracting sector. In fact, the military encourages you to move in that direction. They offer a transition program that is all about transitioning your military skills into civilian skillsets so that you can find a job and transitioning your military benefits into their veteran counterparts.

If you are like most people, you probably haven't really planned for your transition, or given a whole lot of thought to what you want your life to look like after transition. If you are already a veteran, you probably took the Survivor Benefit Plan (SBP) and converted your Service Members' Group Life Insurance (SGLI) to Veterans' Group Life Insurance (VGLI) and wrote your resume. You probably got a job. You have been saving diligently for retirement in a 401(k), Individual Retirement Account (IRA)/ Roth IRA, or Thrift Savings Plan (TSP), and at some point after the age of 59 ½, you plan to stop working and start to draw on those retirement savings, which are intended to provide a stable income for you for the rest of your life. And just like that, a choice that you made to join the military—a choice to be exceptional and to distinguish your life by service—has funneled you directly into the status quo. You have chosen the "smart and the safe" path, and in doing so, you have unknowingly put yourself at the mercy of two massive kinds of risk: psychological risk and financial risk. And unless your life's dream and your sense of purpose in life can only be fulfilled working for a federal contracting company or a brand name corporation, you have just entered the apathy zone.

I'm going to zoom in on the psychological and financial risks of apathy in Chapter 3. But first I want to take a closer look at that turning point I was just talking about: military transition. The choices that we make at transition are largely dependent on the choices that we made long before transition, and the circumstances of our lives as we approach it. Similarly, what happens to us after we leave the military is largely dependent on what we bring with us out of the military. To understand why transition, as it typically occurs today, is problematic, we must once again take a broader view.

2

What Service-Members and Veterans Are Saying

We are on strike, we, the men of the mind. We are on strike against self-immolation. We are on strike against the creed of unearned rewards and unrewarded duties. We are on strike against the dogma that the pursuit of one's happiness is evil. We are on strike against the doctrine that life is guilt.
~ Ayn Rand, Atlas Shrugged

I am going to turn now to the most recent Blue Star Families Military Family Lifestyle Survey. Blue Star Families, founded in 2009 by military spouses, is an organization with a mission to "empower military families to thrive by connecting them with their civilian neighbors...to create strong communities of support." Every year, Blue Star Families conducts a survey of service-members and veterans and publishes the results. Their goal in conducting this research is to increase awareness and understanding of trends in military family life and their ramifications for our serving military, our veterans, and the broader American society as a whole. The most recently published survey (2019) includes information provided by over 11,000 active duty service-members, veterans, National Guard, Reserves, and their immediate family members. This survey is currently the largest and most comprehensive lifestyle snapshot of this segment of the American population.

I have been following the Blue Star Families Survey for some years now, and though some highlighted issues change from year to year, certain findings have remained alarmingly the same. Namely, approximately half of the service-members and veterans that take this survey every year consistently report issues of financial stress and depression. Similarly, a substantial percentage of respondents report difficulties around the military transition process.

Financial Stress

Are most service-members coming from a place of stability and financial security in their military jobs and then getting derailed by the great unknown that awaits them in the civilian world? Hardly. As it turns out, a lot of the issues that get exacerbated by the upheaval and uncertainty of transition are already ongoing while service-members are still on active duty.

Blue Star Families respondents indicated that during their time associated with the military, finances was the greatest stressor on their families.[1] Fifty-eight percent of service-members experience "some stress" or "a great deal of stress" about their financial condition.[2] The big takeaway here is that a substantial percentage of military families are struggling financially even before transition, and people who are struggling to meet their day-to-day and month-to-month financial responsibilities do not have the means to implement any longer-term financial plans, which means they probably aren't even making them.

Veteran Suicide Rates Are on the Rise

That same Blue Star Families Report looks at instances of suicidal thoughts while service-members are still on active duty, and finds some even more alarming data. Less than half (44%) of military, veteran, National Guard, and Reserve family respondents who

1 Blue Star Families 2019 Military Family Lifestyle Survey Report Pg. 11

2 Blue Star Families 2019 Military Family Lifestyle Survey Report Pg. 14, Finding 1

have attempted suicide reported seeking help, and of those that did seek help, 40% did not find the services to be helpful.[3] Why would suicidal service-members not seek help for such a huge problem? According to Blue Star Families: "Concerns about potential impacts on a service-member's career is the most common reason for not seeking treatment."[4]

An even more disturbing and related Blue Star finding, confirmed by the VA's Office of Mental Health and Suicide Prevention, is that depression and suicide continue to be an outsized problem in our veteran community.

The VA's Office of Mental Health and Suicide Prevention provides some context for this disturbing trend.[5] In the category of economic disparities, they state that "some veterans report difficulty in transitioning to civilian positions. Their highly developed skills obtained in the military may not translate to higher-level positions in the civilian world." They also correlate "unemployment and poverty with homelessness among veterans." As of January 2017, the U.S. Department of Housing and Urban Development Point-in-Time Count estimated that 40,000 veterans were homeless, with approximately 15,300 on the street or without adequate shelter every night.[6] Feelings of isolation are a risk factor for suicide, as are the presence of health issues like sleep disorders, traumatic brain injury, chronic pain, mental health disorders, and mental illness.[7] The 2019 Blue Star Family Survey confirms this link between military health issues and suicide, reporting that "existing research suggests that 52% of service-members who died by suicide received some form of

3 Blue Star Families 2019 Military Family Lifestyle Survey Report Pg. 24

4 Blue Star Families 2019 Military Family Lifestyle Survey Report Pg. 12, Finding 6

5 2019 National Veteran Suicide Prevention Annual Report, Office of Mental Health and Suicide Prevention, U.S. Department of Veteran Affairs

6 2019 National Veteran Suicide Prevention Annual Report, Office of Mental Health and Suicide Prevention, U.S. Department of Veteran Affairs, Page 6

7 2019 National Veteran Suicide Prevention Annual Report, Office of Mental Health and Suicide Prevention, U.S. Department of Veteran Affairs, Page 7

health care (not necessarily suicide-related health care) within 90 days prior to death."[8]

Clearly there is a problem here. Depression, mental illness, and suicide are complex, multi-faceted issues. Every suicide is the culmination of a unique set of personal circumstances. But if there is any generalization that can be made about suicide, it is this: people who feel like they have something to live for don't kill themselves.

The Military Transition Program

So now that we've established that a substantial number of service-members that go through the military transition program are bringing existing problems into the program with them, let's see what service-members and veterans have to say about their experiences with transition itself. Trends continue to show that veterans do not feel transition assistance programs are preparing them for a successful transition into civilian life.[9]

As part of their work to help "service-members, veterans, spouses and employers better understand and prepare for the military-to-civilian transition process and civilian employment," an organization called military-transition.org collects information from veterans about their transition experiences. As of the time of writing this book (April 2020), their survey results indicate that approximately half (48%) of the veterans who took their survey found their transition to be more difficult than they expected. This survey includes an opportunity to leave comments. As of April 2020, they have collected 1,064 comments from individuals across all branches of service, and collectively, the over 500 comments left by veterans who found their transition to be more difficult than expected paint a disturbing picture. I have included just a handful here below:[10]

---

8 Blue Star Families' 2019 Military Family Lifestyle Survey Report Pg. 24

9 Blue Star Families' 2019 Military Family Lifestyle Survey Report, Finding 6

10 https://www.military-transition.org/comments-2.html

- (I) feel the transition was not at all useful.
- I had absolutely no idea how to do it and no one to help or answer questions.
- Like many of the false promises made by the military, I assumed, erroneously, that I would actually have assistance in my transition and the programs would be in place to help facilitate my transition. None of this was true at the time and probably not true now either.
- Transition best practices and lessons learned are often lost with each transitioning veteran. The veteran community must come together to have a candid discussion on transition. Often, vets are left trying to figure it out on their own with no clear starting point.
- I received no help at all. You...are totally on your own. Transition help is nothing more than a Band-Aid for a bullet hole. America needs a culture change. Until wounded warriors are protected, no service-member has a chance.
- No jobs that pay well enough. If you can't afford to maintain your previous lifestyle everything starts to decline. Living situation, mental health, everything.

As it has in previous years, Blue Star Families continues to report that a majority of service-members experience difficulties with transition. Half of male veterans that responded to the 2019 survey (51%) reported difficulties. Even more disturbing is that 66% of female veterans reported being "less prepared on every aspect of transition (i.e., overall transition, employment, higher education, navigation of resources in the local community, and navigation of the health care and benefits system)." They also reported having "significantly greater feelings of social isolation." Furthermore, the survey indicates that these problems don't end when "transition" is complete. Instead, they tend to have "long-term impacts on social isolation, financial stability, and navigation of community resources."[11]

The reality is that for many transitioning service-members, the situation is dire, and it doesn't seem to be getting any better, year to year. But let's get real: it's not the military's job to help service-members transition. It's the military's job to win our nation's wars, not to guarantee you get a job. Why on earth would we expect them to be good at basically becoming a job-placement program? Even more important—why would you put your future at risk by leaving it up to them? Notice how cynical many of those quotes were about transition. These individuals took no responsibility for their futures. Instead, they blamed the military for not doing a better job of teaching them how to be a civilian. WTF?

What Does It All Mean?

For some, military service is a difficult time and full of disruptions. Their service comes along with financial instability, difficulties in educating and caring for children, and perhaps even, in the case of unemployed or underemployed spouses, dissent among the ranks. If this is where service-members are coming from, it's no wonder that so many of them struggle through transition and choose to take the "smart and safe path" when they get out of the military. The lure of a stable corporate salary and an eventual retirement pension must be irresistible. But is it all it's cracked up to be? Maybe. Maybe not. I'm going to talk some about that.

Regardless of what path you have chosen or ultimately choose, the bottom line is that you must take 100% responsibility for what happens to you. The moment you give any of that to someone or something else, then you lose control and set yourself up for disappointment.

In the last chapter, I mentioned that there are both psychological and financial risks inherent in the "smart and safe path." In the next chapter, we're going to start naming those

11 Blue Star Families' 2019 Military Family Lifestyle Survey Report Pg. 9

risks and take a look at what they mean for service-members and veterans in the long run.

3

The "Smart and Safe Path"

The simple step of a courageous individual is not to take part in the lie. One word of truth outweighs the world.
~ Aleksandr I. Solzhenitsyn

For the transitioning service-member, the "smart and safe path" begins with the traditional military transition program itself. Transitioning out of the military can be a mixed bag of emotions. There is excitement, of course, to be starting a new phase of life. Veterans can finally do all the things that they have dreamed about doing, and deferred doing, while they served their country. But it can also be a scary time. As we saw in the last chapter, even with a steady military paycheck, roughly half of a representative sample of active duty military are already experiencing some form or combination of family, financial, and emotional/psychological stress within the structure of the military. Now they are going to be on their own: no more orders.

To compound the problem, many transitioning service-members don't understand their own finances. How do I know this? Because I have seen it first-hand, working in the financial industry serving service-members and veterans almost exclusively since 2008. At the transition point, most service-members have an immediate need: to reestablish structure and financial stability in their lives, to replace that paycheck, hopefully with a bigger one.

Because so many do not have a plan for what they want to do next or the resources to spend some time post-military figuring it out, they tend to look only at the surface of the problem. This failure to connect their financial decisions to what is most important to them in their lives can lead to impulsive financial decisions they may later regret.

Before I start talking specifically about finances, though, I want to pause here to make the point that some service-members compound the problem of transitioning out of the military by putting off making a decision about what they are actually going to do post-military. Some people truly have a passion for military service, and the military is their chosen career. But many career military don't necessarily feel that way; they continue to advance in rank and years of service because they recognize that the military is a "good job." They don't know what they want to do when they get out of the military, so they simply stay there. They extend and extend and extend. Maybe they tell themselves that it's to get that next promotion and the pension bump that comes along with it. After a while, remaining in the military, then, becomes not the pursuit of a passion so much as an avoidance of the need to make a decision about what comes next. Separation from the military is coming at some point, though, one way or the other.

Then there are those who know that it is time for them to get out of the military. They have had what I called in my last book, *Life, Liberty, and Purpose,* the "ah-ha moment," but they are no closer to knowing what should come next for them than the service-members who keep extending. Some of them decide to use the GI Bill to go to college. Some join the Reserves. There is nothing wrong with either of these options, if you are exercising them with intention and purpose. Unfortunately, the intention and purpose with which many veterans pursue these options is simply to avoid the need to decide what to do next in their lives. Acting with a lack of direction and purpose can work out on the surface. You can certainly keep yourself and a family fed and sheltered that way, and you can even end up with the common external markers of success. But as I will explain in the next chapter, if the things that you do in your daily life are not aligned to some deeper meaning or purpose that is particular to you, apathy is inevitable.

The GI Bill

If you visit the GI Bill page of the Department of Veterans Affairs (VA) website, the educational future of our country's veterans looks pretty bright. The VA tells us that since the post-9/11 GI Bill came into effect on August 1st, 2009, it has benefited 773,000 veterans and their family members. Does that translate to 773,000 happy and fulfilled veterans and veteran family members living purpose-driven lives?

Um....no.

Actually, the Department of Defense, the Department of Education, and the VA have reported that very few veterans who use the GI Bill actually finish those degrees.[1] The sad truth is that all too often, some of these veterans are really only using the GI Bill just to get the housing allowance that goes along with it.

How do I know this? Because that's what many of my friends did. I, myself, have post 9/11 GI Bill benefits, even though the taxpayers already paid for my West Point degree, because I served for 18 months past my initial commitment. I could have enrolled in classes to get the housing allowance when I was lost and broke, but I hated school, so I didn't. I also didn't feel good about using tax-payer dollars that way. And I didn't see the point: going to college would only "matter" if I finished the degree, and I knew I was too lazy to pull that off. Instead, once I had my ah-ha moment and recognized my own apathy, I used MY money to

1 https://bit.ly/student-veteran-graduation-rate

https://bit.ly/post-911-GI-Bill-Usage

NVEST-Report at https://nvest.studentveterans.org

invest in my education by taking online courses from people who had achieved real success in the real world. It was much quicker than going to college and way more valuable. I now have real online marketing and branding skills that no college could have taught me.

And what about those veterans that take the GI bill and do finish college? Some veterans do earn their undergraduate or maybe some advanced degree. This would be great if it was being done with purpose and intention as part of a larger plan to pursue some calling. But judging from what veterans are telling us about their post-military experiences, apparently that's not what's going on. A lot of veterans earn MBAs, thinking that these letters will add some value to their resumes and somehow magically make them qualified to do something. They don't. Some of these guys optimistically take their shiny new degrees to job fairs, but they don't know what kind of job they want. Neither do they know what they're qualified to do or what value they have to offer a company. This makes them no better off than before they spent all that time and the public's money on their degree.

To illustrate the point, let me tell you a quick "rich veteran friend, poor veteran friend" story. My "poor" friend Mitt used the GI Bill to get an MBA. A decade later, he's barely ever held a job. He's never used his MBA once. But he looks at what I am doing with my life and accuses me of being a snake oil salesman. My "rich" friend Chris Berisford, on the other hand, didn't use any GI bill. Instead, he invested his own time and money to learn about real estate. Now he's not only a successful real estate investor, he's a top real estate agent; you know, the kind that actually helps clients. (If you're in Colorado, ask me for an intro to Chris.) Do I have rich friends who did take the GI Bill? I sure do. I know some veterans who ended up on Wall Street making big bucks. They look like amazing success stories, from the outside, but Wall Street made them miserable. I will only touch here upon a point that I will return to later: riches and wealth are not the same thing. This book is not about how to become "rich" instead of "poor." This book is about how to become "wealthy" instead of "apathetic."

I'm not putting down the GI Bill. It's a fantastic tool for someone who knows what they want to do and the steps they need to take to get there, when a college degree is one of those steps. Unfortunately, that's rarely what's going on with veterans who use it.

The big takeaway here? Education and college are *not* synonyms. If you are using or plan to use the GI Bill, take a real hard look at your reasons for doing so. Are they purpose-driven? If not, they may be expressions of apathy. More on that in the next chapter.

The Reserves

Some veterans stay in the Reserves to preserve close contact with the military, get a little extra paycheck, and make some connections. A lot of self-employed guys will do that. Some veterans are able to combine a start-up business with the Reserves, like I did, using the Reserves like a side hustle to help make ends meet. What, you may wonder, could be wrong with that?

I mention the Reserves here because like the GI Bill, it's all what you make of it. Many of the Reservists I served with took it much more seriously than I did. It didn't take me long to start feeling that there was no value for me in being in the Reserves just to get my "retirement points," and eventually I just stopped going. I didn't feel good about getting paid that way. So I speak from experience when I say that joining the Reserves after active duty can be a crutch, a way to cling to an old identity while avoiding finding a new one. It can function as an obstacle, making veterans "too busy" to uncover and pursue their real passions. If you have or planned to join the Reserves, are you acting with purpose and intention, or are you procrastinating in moving on with your life? Again, the key is to know why you are doing what you're doing.

The *why* is everything. I dedicate a whole chapter, in the second part of this book, to knowing your *why*.

But most separating service-members don't take the GI Bill or join the Reserves. The vast majority simply go through the standard military transition process. There are two aspects to this, both of which can be looked at from a financial point of view and a psychological point of view. The transition from military to civilian life is a minefield of financial risk. The decisions that you are asked to make during the transition process have extremely long-term repercussions, and unfortunately most transitioning service-members make these decisions with little or no understanding of what they are signing up for or what those risks are. There is the aspect of replacing your military occupation with a civilian occupation (translating your skill set, writing your resume, and looking for a job) and replacing your military benefits with civilian alternatives (whether or not to take the SBP and whether or not to convert SGLI to VGLI). Psychologically, there is the aspect of replacing your military identity with a civilian identity, which is also an aspect of looking for that civilian job. For the rest of this chapter, I am going to talk about the financial risks inherent in the status quo alternatives. Then, in the next chapter, I will dive into the psychological risks of these options and how they lead so many of our country's greatest resources and next would-be community leaders into apathy.

Replacing Military Occupations with Civilian Ones

In the standard military transition process, the military provides you with strategies and guidance intended to help you translate your military skills into skills that the civilian world understands so that you can replace your military paycheck with a civilian one. The "translation" of skills is obviously a huge issue, judging from the fact that Google has launched a military skills translation tool.[2] It just so happens that this is also what almost all separating service-members think that they should do and have to do, in part because working for someone who in turn takes care of you has been ingrained in their heads by their military service as a way of life.

As I explained in the last chapter, few separating service-members can afford *not* to take this path. Most are living paycheck to paycheck in the military and expect to do so after the military as well. Any "financial control" they may think that they have while they are in the military (the steady paycheck) is really an illusion; you don't have financial control if you are just one missed paycheck away from financial ruin. So separating service-members in this position go through the standard military transition process and follow all the status quo guidance, and most get a job. That much is clear, but that's only the outside of what's going on. What's the real story? What kinds of individual experiences are our veterans having? How does the actual transition go? What did it feel like? How do they feel about it in retrospect?

At military-transition.org, a lot of veterans say not so good:[3]

- As a 22-year Army veteran, I expected my level of leadership experience would work to my benefit. However, the opposite was true...while I had the experience, education, and certifications that employers were seeking, I had no "corporate experience," so I was usually brushed aside.
- With regards to articulating my behavioral fit and work-skills, the hurdles were much higher than anticipated due to my interviewers' stereotypes about "people from the military."
- Even though I was an experienced (11 years active duty) military officer (O-3) with an MBA, every company that

2 https://bit.ly/military-skills-translation-tool
3 https://www.military-transition.org/comments-2.html

showed interest wanted me to start at entry level w/ pay about 35% less than what I was earning in the military.

- I believed my 22 years of service meant it would be an easy transition, but I learned very quickly that a lot of employers don't understand service-members. They don't equate our experience as real work experience. Salaries are not matching the jobs...
- I transitioned to a new location after working in two different jobs over a 3-year period. There seems to be no incentive for companies to hire more senior people, especially when making a major career change. I also had a limited network because of the move. The applications I did online rarely got a response back. Interviews were limited. Ended up taking a job at XYZ company to get busy, get experience, and to show that I am willing to take an lower-level position with the potential to grow into more responsibility.
- Well, the military tells you you are a wanted employee, civilian sector pays more, etc. Well, I found none of this to be true!
- You have to set your expectations. I was a senior non-commissioned officer, had a masters degree, had a PMP certification, and thought my God, getting the next job is going to be easy...it wasn't.
- Given my experiences, skills, and civilian education (all fully transferable and coveted by multiple industries), my transition should not have been as difficult as it was. I certainly didn't have unreasonable needs.
- I applied to 306 positions and got one solid offer, which I took out of desperation.
- I expected it to be easier. You hear stories of officers getting out of the army and making $200K+. That is simply not the reality. Veterans getting out need to realize that they may even take an initial pay cut when they get out. When

you get out of the military, you are starting a new career—you'll start at the bottom of your field and need to work your way up.

- There are hundreds of comments like this at military-transition.org, but you get the point.

In the last chapter, I told you that according to the 2019 Blue Star Families survey, respondents indicated that during their time associated with the military, the greatest stressor on their families was financial issues. But that's only half the story. The other half is that apparently financial stressors increase after service-members leave active duty.[4] Money-related issues are the greatest stress for 44% of service-members and 49% of service-member spouses, while they are apparently the greatest stressor for 53% of veterans and 53% of veteran spouses. The corporate world offers a paycheck, (supposed) application of a skillset, and a set of benefits, but what you actually get (when you finally find a job) is:

- Limited income potential (your income is capped)
- A middle-class lifestyle that probably includes debt
- Stuck in a skillset and body of knowledge which may or may not keep up with what is going on in the rest of the world, or stuck applying the same very narrow set of skills and knowledge over and over again
- Lack of financial control (your money is tied up in traditional retirement plans and paying for benefits)

I'm going to talk about the psychological risks of taking this route in detail in the next chapter. But first, let's take a look at the other financial aspect of the military transition process: replacing military benefits with their civilian alternatives.

Replacing Military Benefits with Civilian Ones

I have written at length about the pros and cons of the SBP and VGLI on my website, and about the unique alternative solution

4 Blue Star Families 2019 Military Family Lifestyle Survey Report Pg. 11

to them that I have designed. If you aren't already informed about these plans, I highly recommend that you read my e-book and watch the webinar on the subject, which you will find at SurvivorLibertyPlan.com. While I am not going to get too down in the weeds on either one of these programs in this chapter, I do just want to touch here on the risks inherent in both of them. I will talk about the risks inherent in some of the better-known alternatives to them, term and traditional whole life insurance, in Chapter 15 when I talk about asset protection.

The Problems with the Survivor Benefit Plan

Whether or not to take the SBP is a one-time decision that must be made at retirement, or within a year of a change in life circumstances, like remarriage or parenthood. If the veteran has a spouse at retirement, then it isn't even a choice. The spouse and children are automatically enrolled at full coverage (55% of the pension) unless the spouse elects a lower amount or declines coverage. The decision as to whether or not to take the SBP rests *solely* with the spouse, who must opt out with a notarized signature. There is no requirement to qualify during the underwriting process. There is also no requirement to set up premium payments, as the government deducts the money directly from the veteran's pension.

Sounds pretty good, right?

There are circumstances in which the SBP is a lifesaver. But for many veterans, the devil is in the details:

- There is no requirement to qualify during the underwriting process. This is presented as an advantage, but in reality, it is a drawback. A plan that doesn't have qualification requirements isn't an advantage for someone in excellent health, who could qualify for a much lower premium rate in the private marketplace.
- It is expensive, especially given the low likelihood that a retiree's spouse (or a child under 21) will ever actually receive any tangible benefit from it. There is zero possibility

that anyone other than a spouse or a minority child will receive the benefit.

- No benefits will be paid (and no refunds given) if the spouse predeceases the service-member. This is worth considering in cases where there is a significant age gap, the spouse is ill, or the service-member is female (statistically, women live longer than men).
- The SBP does not offer any equity or return on investment while the insured retiree is still living. It only pays out if the insured retiree dies.
- Children over 21 cannot be beneficiaries. If the spouse dies at any point after the retiree, if all children are over the age of 21, SBP payments stop.
- It is complicated to manage during major life events like divorce, and there is a risk of losing the benefit.
- If the insured retiree doesn't die within the 30-year term of the payment plan, then there is a significant (hundreds of thousands of dollars) opportunity cost.
- The processing time to receive SBP benefits is significantly longer than the processing time for other life insurance products. It takes a minimum of 45 to 60 days to start receiving benefits, assuming there are no problems handling the paperwork. Other life insurance products will send you a check for the full death benefit tax-free within a week or two.[5]

In the long run, the amount of money the beneficiary receives from the SBP annuity is usually considerably less than what the monthly premium amounts could have generated as investments and insurance in the private marketplace. The *only* way that the SBP makes good financial sense with regards to ROI is if a service-member dies within a few years of retiring. At an annual cost of 6.5% of his/her pension adjusted for inflation, deducted over a couple of years, in return for 55% of his/her pension, again adjusted for inflation, to be paid to the spouse for the rest of the

5 https://www.military.com/benefits/survivor-benefits/survivor-benefit-plan-faqs.html

spouse's life, the SBP is a pretty good deal, as long as the insured retiree dies within a few years of retiring from the military.

The Problems with VGLI

VGLI, administered by the Veterans Administration, has three main problems:

- Like the SBP, there is no qualification required if coverage is accepted within 240 days of service. This is a good option for the service-member with a life-threatening disability who is thus unable to qualify for privatized life insurance. But it's not necessarily the best option for a veteran who is still young and in good health, who could likely qualify for a higher death benefit at a lower rate in the private marketplace.
- Like SGLI, VGLI is the same fixed cost for everybody. VGLI is more expensive than SGLI from the beginning. But many of our veterans without life-threatening disabilities who are paying into this program are not fully aware that VGLI gets even more expensive every five years.[6] What this means is that the older the insured gets, and the more likely they are to die, the more astronomical ($1,840 a month for those 75 and over) the costs become. It's unrealistic to expect that any veteran who's living off their pension will be able to keep up with these costs.
- The death benefit for VGLI is the same for everybody, too: $400,000 max. That seems like a lot of money, but when you consider that the insured is likely to be earning a salary of something around $100,000 a year by the time they die, it becomes clear that $400,000 isn't going to take the widowed spouse and any children very far. Furthermore, if the insured is forced to stop coverage because they are no longer able to afford the premiums, then that $400,000 death benefit is not going to be around at all when the service-member or veteran wants to leave a legacy, even

6 https://www.benefits.va.gov/insurance/vgli_rates_new.asp

if they have been paying into the program for 20+ years, because there is no equity in VGLI.

With so many veterans going through transition already financially stressed, earning less than they expected to be earning in civilian positions, and paying so much more for their benefits than they were paying in the military, is it any wonder that our veterans report an even higher level of financial stress than our active duty service-members?

Before moving on to the next chapter, I want to circle back to that corporate job we were talking about earlier, and touch on one last financial risk that typically comes along with it: the traditional retirement plan that so many veterans seek as part of a corporate benefit package when they look for a job. Why traditional retirement plans in particular? Because as it turns out, status quo retirement planning, as one of the "smart and safe" things to do, much like taking the SBP and VGLI, isn't much smarter or safer than any basic game of chance.

Traditional Retirement Plans

At some point in adulthood, most people are advised to either participate in their company's 401(k) or to set up an IRA. Even the government now does retirement planning this way, as the TSP has replaced part of what used to be a military pension (I will talk more about the TSP and the new Blended Retirement System in a later chapter). With a traditional retirement plan, you elect to have contributions from your taxable income deducted from your paycheck, so you pay less in taxes while you work. The money grows in a securities-based account, and you don't have to pay taxes on any of that growth until you decide to take distributions. Under current law, you are not even allowed to start taking distributions until age 59 ½ without incurring a 10% tax penalty. The prevailing idea is that because we will be earning less income when we retire, our marginal tax rates will be much lower, so paying taxes on the growth at that point won't be so much of a sting.

Since all the funds in 401(k), IRA, and TSP accounts are subject to ordinary income taxation, we have an incentive to keep the money in those accounts and not spend it unless we need to, lest our marginal tax brackets increase based on whatever we distribute. Once you reach the age of 70 ½, however, leaving all of your money in these accounts is no longer an option. At that point, you are required to take what are known as Required Minimum Distributions or (RMDs). So far so good. So what's the problem?

There are two wildcards with regards to these retirement plans: market returns and marginal income tax rates.

- Market Returns: The stock market is generally expected to provide an average 7% to 8% rate of return on investments; however, this is incredibly misleading, since this average takes into account the sudden massive drops and the long recovery periods that are typical of the market. Particularly alarming is the sequence of returns risk. During the last few accumulation years or the first few distribution years on a securities portfolio, the owner is only ONE market correction away from financial ruin.

Marginal Income Tax Rates: The United States was $22.5 trillion in debt at the beginning of 2020, and we added trillions of dollars more during the first few weeks of the coronavirus. Standard and Poor downgraded U.S. credit in 2011, and it has yet to recover. You do not have to be a fiscal or monetary policy expert to understand what that means. When a consumer takes on too much debt and has a reduced credit score, the interest on that debt increases dramatically. That is what is happening on a national level. Uncle Sam can only do two things to pay down this interest: borrow more and increase taxes. You can expect both to happen in the next 10 to 15 years. This will dramatically increase the taxation on IRAs, 401(k)s, and TSPs just as their owners start to take distributions. The fact that the government limits the amount that can be contributed to Roth IRAs is also telling.

To make matters worse, there are ten thousand baby boomers retiring daily. They are now living off of the money that they had invested in the stock market. This is the demographic of the American population that has all the money. Millennials are living paycheck to paycheck and struggling to pay student loans. When the next major crash comes, if we aren't already in it, then who are the investors going to be who will buy at the bottom of the market so that it will rebound? It won't be the baby boomers. They'll be hoarding their money in cash because they have to live off of it now, not forty years from now. They'll be understandably afraid of losing more. The millennials don't have any money to invest. So who is going to have money to help the market rebound? The government, of course. They'll print money out of thin air and bail out the economy just like they did during the 2008 housing bubble crisis; just like they did recently during the coronavirus. The problem just keeps going on and on, and it will keep going on and on, until one day, like what happened to Rome, the barbarians will be at the gates. We will be far too soft to do anything about it, and America as we know it will collapse, just like all empires eventually do. But that's a subject for another book and another day.

Your job for now, and the goal of this book, is to position yourself financially to be able to first, protect your family, and second, try to do something about what is going on in this country. Join me in this movement to regain the liberty that we swore to defend. Unfortunately, I fear our enemies are now mostly domestic. They are the bureaucrats and bankers who think it's their job to control your life. Fighting them in the streets would be a really bad idea, but we can fight them with our wealth and influence. We need more veterans to position themselves financially to play the game and take back control from these enemies instead of continuing to follow their orders.

4

Following Orders Into Apathy

Most people are other people. Their thoughts are someone else's opinions, their lives a mimicry, their passions a quotation.
~ Oscar Wilde

While major life changes such as transitioning out of military service require time, there are financial implications the longer one goes without a plan. Veterans want to be financially secure, but as we saw in the last chapter, most of the financial planning done for post-military life is impulsive. Think about it: transitioning out of the military means that a person's career—their whole life as they have known it, in fact—is coming to a screeching halt. The knee-jerk reaction (and the most impulsive thing that they can do) is to stabilize their situation by taking the first job that they can get. At the same time, they are attending retirement briefings and being required to make serious, long-term financial decisions that most of them don't understand. All too often, those decisions, too, are made impulsively.

The consequences of this are much greater than replacing one paycheck with another and maybe not getting "the best deal" on insurance. Veterans who take this route unknowingly live without a new purpose (when in the military, we were always assigned a purpose). They don't understand their worth as a veteran in

America, and they don't realize that they can have security and continue to serve as a veteran.

The Hidden Components of Apathy

In the introduction, I talked a little bit about apathy as being a lack of interest and enthusiasm. But I want to dig down a little deeper into the concept of apathy right now because lack of interest and enthusiasm are only the manifestations of apathy, they aren't the drivers of it. Apathy is a response to something, and I want to talk about what those somethings are, because they're important.

Lack of Understanding

There are several commonly misunderstood aspects of transitioning out of the military and back to civilian life. The first thing is that most people don't really understand their benefits. The second is that most people don't really understand the modern economy and the fact that today, the "smart and safe path" is neither smart, nor safe. I am going to talk about the modern economy and the new options it has brought us in a later chapter, so I won't go into that here. I gave you an overview of the problems with the SBP, VGLI, and traditional retirement plans in the last chapter in the context of their financial implications. Here, I just want to touch on these programs in slightly different context, which is that service-members and veterans are making these serious financial decisions without really understanding them. The information overwhelm that almost always goes along with transitioning service-members' first real introduction to the government retirement programs is a major component and driver of apathy.

Due to the high op tempo since 9/11 and the many changes to the retirement system, modern military retirees and their families are severely under-educated when it comes to their benefits. The first time many service-members truly begin to learn about and understand their retirement benefits is when they attend one of a firehose of retirement briefings about 12 to 18 months before their separation. The timing of these briefings is problematic in that it removes the option for military families to take ongoing ownership of their financial situations the whole time they are in the military, rather than right before it's time to get out. Of course, no one is stopping them from making financial plans while they are in the military, and the information presented during these retirement briefings is "out there" to be found for anyone who thinks to look for it, but no one does.

Why? For one thing, as we have already seen, about half of active duty military families are already struggling financially just to make it from month-to-month. Note that this is pervasive across all ranks. I've seen many 30-year senior officers who are leaving the military flat broke. It's because they are understandably focused on their immediate needs, not the long term. Second, you don't know what you don't know. There is also an assumption that the "benefits" offered after retirement are going to be comparable, and of comparable costs, to those that service-members enjoy while they are still on active duty. But as we saw in the last chapter, they are not. Military families are not informed how much the costs for these programs will increase upon separation until it is time to sign up for them, making it difficult to make accurate financial plans for transition and beyond.

Let me lay out the numbers real quick. A 45-year-old Army lieutenant colonel (LTC) would be used to paying about $300 per year in government insurance costs (SGLI) while on active duty. Should he die on active duty, all other benefits are 100% subsidized. The day he leaves the military, he's now responsible for paying a far greater share of any government insurance. It's easy math you can look up for yourself on the DFAS and VA websites. That $300 annual cost jumps immediately to $1056 for VGLI and another roughly $4000 (6.5% of the LTC's pension) for SBP! Retirees and their spouses are often shocked when they see the costs involved in both survivor benefit protection and replacing SGLI with VGLI. Spouses (who aren't usually included in these briefings) are shocked to learn how little they'll receive in survivorship payments if the worst actually does occur.

Furthermore, while the government presents these as benefits, they are really not; the service-member and veteran are paying for them. Government insurance packages are still insurance, after all. They are managed by someone who doesn't work for free. SBP, SGLI, and VGLI are only benefits in that coverage is guaranteed without qualification. There is a false perception that SBP is the only option available for ensuring that in the case of a retiree's death, the surviving spouse will continue to receive financial support. Any retiree or spouse interested in alternatives finds that there is almost zero guidance available regarding privatized options on the free market. The result of this confusion and limited information, the prevailing idea that SBP is part of the retirement program, along with a habit of following orders (in some cases for decades), means that roughly 80% of career service-members feel compelled to enroll in SBP and/or VGLI and commit themselves to the program for life. This means millions of dollars of taxpayer-funded benefits are being redirected right back to the government each year to fund existing SBP payments or to pay a third-party contracted insurer (Prudential and the re-insurance companies). Because much like social security, the continuity of the program depends upon each succeeding generation paying into it, the government is forced to entice its employees to enroll rather than encouraging military families to ensure they are doing what is in their best interest.

The truth is, these status quo military financial planning programs are confusing and chaotic, making it difficult to weigh your family's options when trying to decide what's best for your family's unique situation when you retire from the military. Further, the whole transition process is one of information overwhelm, and when faced with information overwhelm, people tend to shut down.

Many transitioning service-members and veterans worry, "Did I make the right financial decisions?" They feel like there was too much information presented to them, when in reality they weren't given enough information. They weren't told that they had other options besides the SBP and VGLI, and if they were, they weren't told all the options. They were only told the not-so-great ones. They don't want to put their families at risk, so they "choose" to accept what the government is offering them out of fear. But you have to have a choice in order to make a choice.

The timing of the presentation of these benefits, which limits the amount of time service-members have to investigate other options; the lack of awareness that most transitioning military have with regards to other options; and the sense of urgency with which the government presents the SBP and VGLI leads inevitably to confusion and self-doubt. This works to the government's advantage, because an individual who is confused and who doesn't feel they can trust their own judgement is primed to take "advice," which is sometimes just another way of doing what you're told. The result? You didn't actively engage in making an informed decision about your financial life because you don't understand it, so you don't feel invested in it. You comfort yourself with the myth that this is how it's always done, and that it will somehow "work out." How could it not work out when you've done the "smart and safe" thing? Yet, you're neither interested in your "choices" or enthusiastic about them: you're apathetic.

Lack of Power

People who have a stake in what is going on are not apathetic. People who can effect change are not apathetic. But people who know that nothing that they do will really make any difference in their situation are apathetic.

If you get up every day and go to a corporate 9-to-5 job, are you powerless? On the one hand, having the ability to meet your responsibilities—to keep a roof over your head and food on the table, to pay your bills, to participate in the consumer economy—is certainly a form of power. But it's power within certain limitations, and there comes a point at which limited power becomes no power. That point is the threshold between the status quo and what lies beyond it. In a traditional corporate 9-to-5 job, you have power only inside your box, and most people don't try to venture outside of that box.

Think about it. A salary is great. It's a guaranteed income, and income is a form of power. But a salary is also a limit. There is no paid overtime when you work for a salary. No matter what you do, you will only make so much money. Whether you take a technical path or a managerial path, step-ups in income are only a few thousand dollars at a time, and there is at least usually a year between them. Even most of those who start a job with genuine enthusiasm, wanting to make a real difference and advance, inevitably settle into doing only what is required because there is no real incentive to do any more. You have almost certainly signed something to the effect that anything that you create or invent, along with any associated credit, patents, or profits belongs to the company, not to you. Your income potential is fixed. Maybe you get a spot bonus of a few thousand dollars every once in a while, an end-of-year bonus, or some kudos published in a company newsletter. But if going above and beyond in effort or ingenuity makes little difference in your overall living situation, why bother? There's a reason that so many government contractors frequently change companies in endless pursuit of a higher paycheck. It's the only real opportunity available to them for making a substantial leap in their quality of life.

If you don't think that this is true, then ask yourself: how much credit card debt do you have? How much mortgage debt do you have? Do you make a car payment every month? Every penny of your debt represents your need and/or desire to exceed the limits within which you have chosen to live. And if you are consistently exceeding the limits of your earning power, then how can you possibly say that you are satisfied with it? Most people see credit as an expression of personal power, but unless you're paying off that bill every month, you shouldn't be looking at your credit as an expression of personal power, you should be looking at your debt as a manifestation of just how much power you lack.

A lack of personal power manifests in your life as apathy. A life of autonomy, on the other hand, is a life in which you do not allow others to establish your limits for you. But we'll get to autonomy in detail a little later in the book.

Identity Crisis

What motivates you? If you are caught up in the resume-writing, job-fair-schmoozing aspects of transition right now, what is it exactly that you are hoping to find? If your honest answer is that you are straight up looking to replace a paycheck, okay, then what? Once you secure the job and have the paycheck, will you be excited to get up every morning and go to that job? Will you be working in a field that interests you deeply and that you feel passionately about? Will you feel fulfilled and satisfied with the contribution you are making to something bigger than you are, of which you feel you are a significant and meaningful part?

You can say what you want about the military—I mean, let's be honest, there's a lot of suck in the military—but as a service-member, you know why you signed up. You know why you are there. You signed up to serve and protect your country. In taking the action to sign up, you were expressing your underlying core values. The individual assortment of core values that service-members have may vary, but they all have at least one in common: they're proud to wear the uniform.

How are you expressing (or how do you plan to express) a passion for service in your post-military life?

If you are like most people, you have probably not given this any thought. You may not even think it is that important right now. After all, it may seem that this particular core value is at odds with other core values that you are expressing through the actions of looking for a job, values like independence, personal and social responsibility, and caring for your family. These are all extremely important values. So important, that it is very easy to both let them overshadow your passion for service and to rationalize doing it. After all, when you are fulfilling your passion for service in a way that you have chosen, in a way that feels right and good to you, then you are in a state of happiness. Choosing to pursue what makes you happy over doing the responsible thing is selfish, right? And if it doesn't work out, if you struggle and fail, then hindsight will inevitably tell you that choosing to pursue your dreams, follow your bliss, was also downright stupid.

Here's the thing. The status quo, that averageness and mediocrity I mentioned at the beginning of this chapter that no one wants to admit to, rarely presents itself as such. It always arrives dressed up in fancy clothes with a convincing argument on its lips. It's not the status quo, it's how it's done. It's not averageness, it's good sense. It's not mediocrity, it's financial responsibility. All of which is to say that the status quo looks like the smart and safe path and produces enough of the trappings of a "successful" and financially secure life to make the argument that it is the "right thing to do" extremely convincing. From the outside, that is. Remember all the debt we just talked about in the previous section?

What no one ever really tells you, the thing that I am telling you, which is what makes my approach to life planning so radical, is that the status quo life doesn't feel on the inside the same way that it looks on the outside.

Now, to be fair, there are some people for whom what they really want to do, what they have a passion for, what excites them and what aligns with their sense of purpose also just happens to align with the status quo way of doing things. People who want to be doctors, lawyers, teachers, or government senior executive service (SES) can choose the status quo path with a real chance at finding fulfillment. But that's only because they haven't chosen the status quo, they've chosen to work in the fields of medicine or law or education or public service. When you choose the status quo only because it is the status quo, and only because you think it is the "right" thing to do; when the status quo path you choose is unconnected to your true interests and passions, over time, you start to experience a kind of identity crisis. The disconnect can be barely perceptible at first, beneath a sense of pride or validation at having gotten a "good" job, beneath the satisfaction of being able to pay your bills and have nice things. But trust me, the gap gets bigger and more uncomfortable as the years go on.

In the military, when we receive a rank and a uniform, we receive an identity. That identity has a cap, just like income does. For most of us, in spite of this cap, our military identity satisfies us enough while we are in the military because it aligns with and is a way to express our passion for service. Similarly, we take on an identity when we get out of the military and get a corporate job. When someone asks you who you are and you respond with, "I'm a (job description with X company)," the question is, does this job description with X company align with and express who you are? Is it a reflection of your personal mission and your passion in life? If not, then in taking this job and using it to define yourself, you have put a cap on your identity.

An identity crisis is a personal psychosocial conflict that has at its core confusion about the social role you are playing and a sense that you are no longer living according to who you feel that you are. If you are in a state of identity crisis, you may feel confusion about the nature or direction of your life. The kind of identity crisis that I am talking about manifests itself as an uneasy sense that you are not in the "right place" or that you are not around the "right people" or that you are not the same person on the outside as you are on the inside. You may feel stuck, unable to make a decision because you "don't know" or you are "confused." This state of unknowing or confusion is often a manifestation of the fact that you *do* know what you want to do, deep down, but you are afraid to actually do it because it doesn't align with what some force or forces outside of yourself (some form of family or social pressure, i.e., the status quo) is telling you is what you should be doing. Perhaps there is no established path for earning a living doing what you really want to do. Perhaps you don't have any guidance or role models, and you don't feel you can justify doing what you want to do as opposed to what you feel you should do. You feel stuck because your desire to do what you want to do (whether what that is has reached your conscious mind yet or not) is just as strong as your compulsion, imposed on you by society, to do the "smart and safe" thing.

If you do the "smart and safe thing," you may begin to feel that you are not living your authentic life so much as you are playing a role that was written for you by someone else. You may begin to wonder if you are the only person that is only "acting." This sense of unease may progress to frustration (I hate my job, I hate my commute, who makes up these stupid rules? I have to remember to buy a lottery ticket...) or rebelliousness (one drink at lunchtime is no big deal...I know I'm not supposed to park here, but screw the stupid rules, I hate this place...) or anger (my boss is such an asshole, I could do his job so much better than he does...). Or your feelings of unease may remain tempered by those other values of independence, personal and social responsibility, and caring for your family, and never progress beyond expressing themselves as apathy.

Gut Check

To circle back, then, to the first chapter: Are you okay with being part of the status quo? Are you struggling financially? Do you feel any lack of understanding, power, or identity in your life? If you want to reclaim your understanding of your finances, your personal power, and your own unique identity, keep reading this book, because I'm going to show you how to do just that.

The Bigger Picture

Putting all of this together with what I talked about in the last chapter, I think it's fair to speculate that our existing transition program is *not* working. A large number of our service-members and veterans are struggling. They are struggling while they are on active duty, they are struggling during transition, and they are struggling as veterans. The problems facing veterans are taking place within the context of problems facing America as a whole. These problems have deep roots, as I discussed in the first chapter. In the meantime, what I want you to take away from this chapter is that the psychological and financial risks of living in apathy inevitably come together to form a problem that can be summed up in this way: most service-members and veterans are currently living within a scarcity mindset.

Having a scarcity mindset means that everywhere you look in life, you see scarcity. For someone with a scarcity mindset, money is difficult to earn, goals are difficult to accomplish, and there is never enough of what you want or need to go around. Individuals with a scarcity mindset tend to resist spending money, even when the return on an investment is significant. A person with a scarcity mindset would rather have a penny today than a dollar tomorrow. And when it comes to your life and the opportunities available to you, if you are looking out at the world with a scarcity mindset, then you are looking out at the world through the bars of a prison, and there isn't enough money in the world to make those bars go away.

A scarcity mindset and apathy go hand in hand.

The standard military transition process centers around money. The reason why it is inadequate is because the problems that veterans are facing are bigger than their finances. I talked about those service-members and veterans who are NOT doing well, but what about those that are? What about the ones that don't feel financial stress and don't have any trouble getting a job when their military service ends? Money can provide the appearance of success. Separating service-members can follow the standard transition process to the letter and by all external measures seem to be successful. They can get a "good job." They can accumulate substantial retirement savings in a traditional retirement savings plan. They can afford to send their kids to college. Many have. But as a financial professional, I know all too well that the "financial security and stability" that the world sees when it looks at their lives from the outside is in fact sometimes nothing more than a house of cards, which can come crashing down at any moment.

The present ongoing economic catastrophe caused by the coronavirus is only the most recent example of how vulnerable too many of us actually are. But finances aside, what I am finding over and over again when I talk to these "successful" veterans is that they *are not happy*.

And if you're not happy, then what's the point?

What All This Means for Veterans Today

Once we took our oath to serve our country, the military kept throwing more paychecks and benefits at us. Then, as we're transitioning out, the idea of service disappears and it all becomes about the money (i.e., paychecks and benefits). Nobody bothers to remember to say: "Hey, I didn't sign up for the paycheck, I signed up to serve the American people," even though this motivation is still inside the hearts of veterans. We face so many distractions and so much noise about the money and the job and the retirement and the kids' college, all of which are important, but none of which are the point of post-military life. But serving, and our passion for service, doesn't end when we take off the uniform.

I'll be blunt: we were taught limited and outdated theories about how to handle our financial futures. No one's paying attention. Veterans are suffering the side effects of financial planning theories and military benefits programs left over from the 70s and 80s. Our guidance has been derived from the least innovative and most corrupt industries led by greedy knuckleheads on Wall Street and K Street. These theories are old and misguided. They are bad for veterans, and they are bad for America.

We need to go back to the original reason why the military was created: to protect the pursuit of liberty; to safeguard freedom and opportunity, innovation and creativity. These are the reasons why we chose to serve. But if we aren't vigilant, the comfort that comes along with the status quo can trump creativity. It can also trump courage. Comfort creates complacency. But most veterans will tell you that they are longing to be courageous for their country once again.

Veterans' understanding of post-military life affects how they treat their personal finances, and their opinion of themselves is hugely influenced by how the rest of the world sees them. The words 'military' and 'transition' can have negative connotations. The government sees the veteran as just another employee. Civilian corporations can sometimes see us as charity cases, and as such, they hire veterans to make themselves look good. There is a pervasive assumption that veterans are poor, and there are countless non-profits out there that want to lend us a helping hand. And how do veterans see themselves when facing transition? As unemployed.

I don't see veterans as another employee, a charity case, or as unemployed. I see them as leaders who are ready to embrace their next life's mission. I see them as a great untapped potential, an amazing resource that this country has that will help us to change things for the better.

Veterans are the less than 1% of the American population who served.[1] They have been given a unique opportunity to be in the 1% of Americans who are wealthy (and who consequently serve many), and they deserve to be in that 1%. It is this viewpoint that made me realize that funneling service-members and veterans into the status quo is morally wrong. Military and veterans are different. We just are. Veterans want to live with purpose, and we need financial guidance that will help us structure our lives with intention so that we can continue to be of service to our country in whatever way we feel passionate about.

What we really need to see is veterans helping veterans. Veterans know how to serve each other best. We are the 1% of the American population that serves, and we deserve to live like the other 1% that also serve the greater good. And we can live like that, if we choose to. *You* can live like that. I am here to tell you that the first step to thriving in post-military life is to let go of the idea that the status quo is the "smart and safe" path. That begins with acknowledging that the "smart and safe" path belongs to the old world. We, however, are living in the modern economy.

1 Americans employed with the U.S. military make up less than 1% of the American population, according to the Defense Manpower Data Center.

5

The Modern Economy

Only those who decline to scramble up the career ladder are interesting as human beings. Nothing is more boring than a man with a career.
~ Aleksandr I. Solzhenitsyn

At this point you may be thinking, okay, so the standard military transition plan isn't ideal, and "just" getting a job to pay the bills has major drawbacks, and I probably will hate it. And maybe my retirement savings plan will tank right before I need to start living off of that money...but what are the alternatives? I have to live. I have bills to pay, a family to take care of (maybe now, maybe later). Eventually, I am going to get old, there's no way around that, and I will still need some kind of income later in life. If everything that everyone else is telling me to do is wrong, then what am I supposed to do instead? What other options should I be entertaining?

Glad you asked.

I am going to go into that in great detail in the second and third parts of this book. But right now, I am going to start this chapter with the story of someone I know who did everything right.

My buddy Pete is a physical therapist. He always wanted to be a physical therapist. The whole time Pete was in the Air Force, he was always saying, "I want to be a physical therapist. That's all I want to do after I get out." When he separated from the Air Force, he took out loans, he bartended, he did whatever it took, and he got precisely the education that he needed. He started making money the day he got that license. Now he's living his dream life and having a blast. He had an interest in a field that he felt passionate about and in which he wanted to be of service, and because that's where he chose to place his focus, he was naturally led to make that dream real. Is he rich? Nope. But he's wealthy, because he controls the source of his income doing something that he loves to do. Today, he enjoys being a PT for the VA helping veterans.

Here's what I want you to take away from this story: Everyone is unique. You are unique. My friend Pete is unique. What Pete did was the *right* thing to do, because what he did was *right for him.* Your task, upon which your own happiness and financial success depends, and the simple and elegant solution to *all* of the risks and problems I have talked about so far, is to find out *what's right for you,* and then *make it happen.*

What Technology and Money Have In Common

Today, people are making money – good money – doing things that did not exist as career options twenty years ago. To talk about the modern economy is to talk about the Internet. In revolutionizing how people communicate and do business, the Internet has revolutionized the world. It has allowed like-minded people to come together over similar interests and goals—repeatedly, daily, hourly, minute-to-minute—in a way that it simply was not possible for people to do when they were constrained by geography. Twenty years ago, the ability to take payments and have them credited to your bank account was tied to brick and mortar businesses. The actual transfer of that digital information required access to a phone line. Setting up the accounts and obtaining the equipment needed to process credit card payments ran into the thousands of dollars and was cost prohibitive for many, and you could only take payments from someone standing right in front of you. Twenty years ago, the ability to promote, market, or advertise an idea or a product was the domain of large corporations and all kinds of publishers who had the power to keep out the little guys and the upstarts as well as the power to decide who they let in. These gatekeepers made and broke careers. Even twenty years ago, your possibilities in life were largely limited by where, and to whom, you were born and where you had the means to physically go.

None of those limits exist anymore. Today, you can connect with and do business with literally anyone in the world. Start-up costs for many things, if there are any, are minimal. Today, you can publish a book and make it available to anyone, anywhere in the world, practically for free. You can sell your arts and crafts. You can turn your hobby into a business. You can do work that you enjoy and that you are good at for clients anywhere in the world. You can network with practically anyone, anywhere. If you want to get a conventional job with a brick and mortar business, you have immediate access to countless people who already work there via LinkedIn, and you can network with those people. You can get educated online. No matter what you want to learn there is someone out there, somewhere, teaching it, and the "somewhere" is no farther away from you than your laptop or your smart phone.

You can even get a college degree online. But do you really need to? Once a college degree was a kind of societal stamp of approval that you were worthy of being given a certain kind of job. Today, you don't have to wait for someone to "give" you a job. Today, you have the ability to create for yourself whatever kind of "job" you want. Even teenagers are becoming millionaires with online businesses. People who would have once been considered on the fringes of mainstream society are finding millions of followers and making millions of dollars on YouTube.

In the modern economy, technologically, anything is possible for anyone.

But technology can never completely get rid of every potential human problem or challenge. Technology no longer holds people back, but people do hold themselves back. Ignorance, lack of understanding, and limiting beliefs absolutely hold people back. Technology cannot change your world view, or make up your mind for you, or get you off of the couch. Technology is an amazing, life-altering thing, but not in and of itself. Technology is a tool. In that way, the Internet is no different from a hammer. A hammer is completely useless until you pick it up and hit a nail with it.

This makes technology a lot like money. Money, too, is a tool. Money is useless unless you do something with it. Money is not, and should never be, an end in itself. When money becomes an end in itself, there is never enough of it ? a destructive point of view called the scarcity mindset that I discussed in the last chapter. Money is just a tool. Money is out there, available for use and waiting to be made. Finances are never a person's true problem. They are the scapegoat. They are the excuse we throw out there between ourselves and the things we want to do.

What then, you may wonder, are the tools of technology and money for?

Technology and money are tools for expressing meaning. If we do not feel that there is meaning in our lives, if we do not understand our meaning, our purpose, our own higher calling, then we have no motivation to pick up the tools and make something with them. We have no motivation to use the tools to create change. And here we have arrived once again back at a state of apathy.

To Begin, Start Searching… for Something…

We often avoid tackling the concept of meaning in our lives. I avoided it for too many years. What changed? I made more money that I ever had in the Army and lost all sense of integrity. Securing a regular paycheck and benefits eventually exposes the true problems and risks in life. Having money inevitably reveals who you are. If there is a disconnect between who you are and what you are about and how you make your money, then once your basic needs are met, once you have no "financial problems" keeping you up at night, that fissure will begin to widen until it becomes a chasm, with you and your money on one side of it and what's really important to you—whatever it is that would best express who you are and what gives life meaning to you—on the other.

We must broaden our approach. We must avoid the false narratives of "that's how it's supposed to be." We must search for alternative ways to build our human capital and discover outlets for our influence. Believe that you have something of value to share because you do. Only you can take the responsibility to do that. Only you can pick up the tools that are laying idly around and do something unique and amazing with them. But before you can see the tools, you have to see yourself. You have to know who you really are. You have to know your own meaning. You have to know your why.

It can be difficult to discover these things, especially if you aren't accustomed to self-reflection. Sometimes it takes hitting rock bottom in your life before you can really begin to see. The journey can be uncomfortable at best, horrifying at worst, because it includes taking responsibility for what and where and who you are in your life right now. That's the hardest part. For years, I blamed senior officers, mentors, coaches, consultants, and online gurus for my unhappiness. In the process of investing in myself and my business, I spent my life savings of at least $500K and change (I stopped counting) searching for an answer to the question: Do I matter?

What I learned is that the root of the problems I talked about in the first two chapters isn't money at all—it's identity.

That discovery is the reason that I wrote *Life, Liberty, and Purpose*. It's the reason I'm writing this book. It's the reason that I founded US VetWealth.

People come to me because they are looking for a guide, someone to give them permission to take responsibility for themselves to figure this shit out. And after having been at the whim of military pay and benefits for so long, they know that settling for advice or passing on the responsibility for controlling the source of their income is a bad idea. They are lost and confused, yet they know there is a massive opportunity happening in the modern economy right now, and they don't want it to pass them by while they're stuck in a cubicle working for Booz Allen or Deloitte.

My Story

You may be wondering how I discovered this connection between identity and money. You may be wondering what the modern economy has to do with it. Or maybe you can't get past wondering what particular variety of unhappiness would motivate someone to spend $500,000 trying to find himself. All fair questions, so here's my story.

Like many of you, I have had a number of identities in my adult life. Before the Army, I was captain of my high school state championship lacrosse team. At West Point, I was a cadet who didn't really buy into all the discipline but flew under the radar. In the Army, I was a Knowledge Management Officer at a major command (EUCOM). Before that, I was a Ground Liaison Officer for the Air Force's 52nd Fighter Wing, and before that I was a Battery XO and a Platoon Leader in the 1st Infantry Division. These are standard military identities, and they are easy to latch onto. After the military, for about seven years, I felt that my identity was that of a West Point grad and financial planner. The financial industry wasn't so much a career choice as it was the first viable thing that I ran into with my then gotta-replace-that-paycheck mentality.

In 2008, when it came time to choose whether or not I was going to stay in the military, I was stationed in Stuttgart, Germany. My next duty station was set to be Fort Sill, Oklahoma. At the time, I was pretty serious about dating a German opera singer. There wasn't going to be any opera singing in Fort Sill, and I had bought property in Germany, so I decided to get out of the Army and stay there. It was a risky thing to do, I guess, and I ended up sacrificing a total of 11 years between my time on active duty and then afterwards that I could have been closer to my family. But I wanted to stay in Germany, so that's what I did. I had no plan whatsoever, and I ended up working in the financial planning industry.

I went and bought a bunch of suits because I thought that's what you're supposed to do when you get a new job. The guy I was working with had been a Certified Financial Planner (CFP) for about thirty years, and he was making at least a few hundred thousand dollars a year. He seemed like he knew what he was doing, and he had hundreds of clients. I shared and helped pay for an office with him with the understanding that I wasn't going to make any money whatsoever until I began to get clients of my own. No problem. I had some money in savings, and I felt confident.

He told me he would show me the industry, but all he ever did was call his own clients, while I sat in the office looking at financial planning websites. This is *not* how you get clients, but it did lead me down some Internet rabbit holes that broadened my perspective. I started to see things differently than what I was being told. It took me years to put it together, as you will see, but I guess my journey really started when I finally realized that this CFP with 30 years of experience had no idea about how the stock market really worked. He didn't understand personal branding or how to create a unique value proposition to attract clients.

He had gotten lucky in the 90s and gotten tons of clients when everyone wanted to invest. He had managed to keep a lot of them and just build their assets and take his fee.

When the 2008 crash happened, though, right about the time I entered the industry, he started losing clients like crazy, and he was freaking out. He even started asking me for advice! WTF? I worked with him until around the fall of 2010, when I finally decided that he really didn't have much to teach me, and I needed to figure things out for myself. So one day I found a new office just outside of Patch Barracks in Vaihingen outside of Stuttgart. One weekend, I went into the office we shared and moved my stuff out. I wanted to do things myself. If I was going to fail in the industry, it would be my fault. And thank God I made that change. Within a year, I had 10Xed my income.

A Word About the CFP

The title CFP tends to give people confidence. It's supposed to. It's the word "certified." Something that has been certified is officially recognized as having certain qualifications or meeting certain standards. But standards for what? Who sets these standards? And in whose best interests?

Contrary to what you probably believe, the CFP designation wasn't created to protect clients. It's really just a marketing schtick. A group of people got together and formed a "board" (which sounds very professional and official), and said, "Hey, our industry is getting a bad rap because of all these stockbrokers, these Wolf-of-Wall-Street-type guys, who are just scamming people. So we're going to create these industry standards, and set up these rules where you have to qualify in order to be 'certified' as a financial planner, and once you get certified you can only do things in a certain way." So where does the marketing come in? Well, when they created the CFP designation they also created a mindset shift, right? If there is a "certification" out there, people tend to consider those who have that certification to be more legitimate and more knowledgeable about their field than those who don't. Now, this may or may not be true. It's certainly not necessarily true, but it's perceived to be true, and as such, it kind of sets a snowball effect into motion. People think that a CFP can tell them the right things to do with their money, and anyone who is or wants to be a financial advisor at that point is then under a certain amount of social pressure to become a CFP. The CFP became the status quo of the financial industry. And remember what we've said about the status quo: it is average, mediocre. The status quo doesn't benefit YOU, the person doing the status quo things. It benefits the people who are pulling the strings on those status quo things, the people who set those status quo things into motion in order to ultimately benefit themselves.

So now you have all these financial advisors thinking that if they get the CFP certification they will easily attract clients, and they'll know to guide their clients to do all the "right things," but what actually ended up happening is that now everybody who goes to a CFP gets the same crappy advice. The CFP became just another commodity, and there's one on every corner.

I looked very intently at getting the CFP a few times, and thank God I never did, because if I had, I would be stuck in this fee-only financial planning model, charging people a $400 monthly retainer (in addition to a $2,500 up-front fee, like some of the financial consulting folks I know), and for what? So I can call them up every once in awhile and say, "Let's adjust your allocation a little bit." Artificial intelligence can do that now. That's why I use an investment platform that offers artificial intelligence. Not only is it going to make better (non-emotional and non-prejudicial) decisions, it also keeps fees down.

Of course I charge a fee, because I'm in business. But I'm not charging fees for investment advice. I charge fees for giving real financial advice about how to create wealth and opportunity for yourself. For example by teaching you how you use LinkedIn as the awesome and amazing networking tool that it is. You've heard it before, your network equals your net worth. My approach to "financial advice" shows you how to get control over your life by doing better things with your money. I help my clients avoid losing money to taxes and to the unpredictability and potential volatility of the stock market. I have no idea what asset allocation is going to get you a return in the market, and guess what? Neither does anybody else. The CFP doesn't come with a crystal ball.

So, that said, back to my story.

Life as a Financial Advisor

Maybe part of the reason that I struggled during that time was that I wasn't motivated by money. I was trying to find a purpose. I wanted to feel like I was actually adding value to the industry and to the people that I dealt with. I was trying to figure out the value part.

In February of 2011, now settled into a new office of my own, I saw the top advisor in our firm, who wasn't a CFP, give a presentation about how he does this thing called holistic financial planning. It was a nice presentation. It was drawn up and explained very simply and nicely. That was the kind of thing I was looking for, because I like getting up and presenting. So I introduced myself to him, and he offered to kind of mentor me. He basically said, "Hey, bring people in, and I'll work with them, and any money we make, we can split it." That made a lot of sense to me. He always used to laugh about making me this offer, because he didn't need any more money. He was the number-one guy in the firm, making hundreds of thousands of dollars a year and living a dream lifestyle in Italy. But he took me on and became my mentor, or so I thought.

To find prospects to bring to him, I would have to do horribly embarrassing marketing things like pay money to sponsor a golf tournament, and I'd stand out there at a booth and awkwardly make conversation: "Hey, Mr. Johnson, hey, how're you hitting them today? Do you want to come over and sign up for my free giveaway?" That's how I would get people's phone numbers to call. Cringe. All together, I spent an odd four years with him. At the same time, my relationship with the opera singer was starting to unravel.

In the spring of 2011, I worked yet another of these sponsorship events. I stood at my booth twelve hours a day for three days at this bazaar in Germany. Everyone was coming out to buy all this crap. It was just absolute consumption. I couldn't believe the amount of money that was being wasted simply because people were bored. How could you be bored while living in the middle of Europe? Most military families spent their weekends in the PX shopping. So I worked the booth at this bazaar, sitting there right next to USAA. Those guys were getting hundreds of people stopping at their booth, and I thought I wasn't doing too well. But I'm a self-taught kind of guy, a hands-on sort of learner, and I'm willing to try new things, so the next time I worked this kind of event, I came prepared. What I did was buy hundreds of colorful balloons, and I had this printed on them: 529. That was to indicate that maybe people should give more thought to what their kids' futures are really going to be about when they're saving for them instead of just putting money in 529 college savings plans.

My thought was that since I wasn't really expecting anyone to come by my booth anyway, I would just sit there and pump up these balloons. I put them on sticks, and I would hand them out to kids as they walked by. Well, what happened was that as the kids with the balloons were walking about the bazaar, other kids started seeing these balloons. "Mommy, Mommy, Mommy, I want to balloon!" "Hey, man, where'd you get the balloon?" So the parents would come over to my booth. Boom. Cool. And I would get an opportunity to talk to them and ask, "Hey, by the way, would you like to sign up for a giveaway?" Who doesn't want a hundred free bucks? So they would sign up for the giveaway, and I would chat them up a little bit. Because of my balloon tactic, I collected about 200 names, and I ended up with a few leads. I was never aggressive. I would get all these names and barely call them because I was so nervous. What would usually happen is that I would run into people that I already kind of knew. They would say, "Oh, this is what you do," and then we would talk a little bit about it, and these people I tended to follow up with.

My Identity Crisis

I was nervous, but I had these 200 names, and I started to get motivated because this new mentor did things a little differently than the other financial advisors I had worked with. I started calling all these names that I got at the bazaar, and the next thing I knew, my mentor was coming in and meeting with people, and he started making sales. Then, basically, I was just sitting there and watching the show. He was good at getting clients and developing relationships. We'd have these three-hour-long meetings, and I was supposed to just sit there and take copious notes, and I'm not good at that.

Now I know better than to waste an ounce of my time doing something that I don't like and that I'm not good at. Now I know I should have outsourced it. It would have been better for me to literally hire somebody to sit there and take notes so I could have been more engaged. I could have gotten ahead way faster. But the way it actually went was that I would sit there and watch. Then I would have to type up all the notes and follow up with people. I was really just the admin guy. But I was also doing all the marketing and bringing in the leads. My mentor would basically just come in, do the presentations, and close the deals. We ended up doing really well. In that first year that we worked together, which was my third in the industry, I went from making no money to making well over six figures.

But this isn't a feel-good, success story. Remember what I said in the last chapter about how money reveals who you really are?

Right before all these paychecks started coming in, in the fall of 2011, the opera singer decided she had had enough of me. Apparently, I didn't like opera enough. By Christmas, I had nothing tying me down financially, and all of a sudden, I decided that I was going to Thailand. I just threw it on a credit card and paid it off. My mentor was down with it. He was a globe traveler as well. He was older and had a family, so they would always go on trips as a family.

I went backpacking. I went to run with the bulls. I went to Brazil for the World Cup. I motorcycled across Vietnam. That's the kind of stuff I was doing. In my decade living in Europe, I managed to travel to about 50 countries and go off the beaten path. I had to learn customs, cultures, and languages on the fly. I had to take major risks with no idea what was happening. I was developing skills I had no idea I would find *extremely* valuable later in life as I dove into entrepreneurship.

On the other hand, during that time, most of my friends ended up moving to new duty stations, or getting married, or getting out of the military, having kids, and starting a life, while I was still just kind of stuck over in Germany. I literally got stuck because I didn't know how to leave. I didn't have anyone to give me PCS orders. It was shocking how little power I felt I had to make a major life choice like that without the government or a "job" telling me what to do and where to live. I owned property, and I didn't know how to rent it out or what to do. And I was making good money. I had an office. I had this relationship with my mentor down in Italy. There were no orders to follow. I went on like this for three years.

Then, one day, in Koh Phi Phi, Thailand, I got a wake-up call. I was sitting on the beach looking at my iPhone (which was a new thing then, by the way, and they already had Wi-Fi there), and I got an email from a joint client I had with the first CFP whose office I had shared. The client had bought a life insurance policy from us because he wanted to help me out. He couldn't afford the policy anymore, and he wanted to make some changes. I was communicating with him on my phone and freaking out because I didn't want him to find out that I was in Thailand. It was the first time I felt like I had to hide my lifestyle.

As time went on, this kind of thing happened over and over again. I did what everyone else did and posted my travel photos on Facebook, and I'd get these stinking comments: "Do you even work, bro?" Or from family members, "I wish I could travel, but I haaavvvvve a job." WTF did these people think I was doing? These are the same type of people who are confused about why my business is booming right now when they and their friends are getting laid off.

I was starting to struggle with my identity because I wasn't doing what I knew I was "supposed" to be doing, which was going into the office and sitting there for eight hours a day, filling the calendar with client meetings. That was the role I thought I had to play if I wanted to be successful in the industry. And that's what my clients were doing ? they were working really hard. But I felt like I had made enough money, and I wasn't motivated to make more. I was really just kind of miserable and depressed. I wanted to get the heck out of Europe. So I didn't do much. I just kind of sat around and answered emails and calls as needed. I wouldn't go look for new business until my mentor in Italy would call and say, "Hey, I'm going to come up for a weekend and meet with everybody." Then I would go crazy, getting a bunch of people on the calendar. We did that every couple of months. And there'd be some follow-ups and some business done. I would do another sponsorship and collect more people that I could call, and that was enough.

Except that it wasn't.

There was no system, order, or business model to what he was doing, and I was just blindly following him. I would ask him, what's the plan? What's the model? His answer was just, "Go meet nice people." What??? He was oblivious to the fact that he was simply one of those extroverts who is confident and able to attract enough attention to fill his ego, but he had no true skills to pass on. He couldn't have told me the definition of branding, let alone how to build one. In the traditional financial industry, though, if you're like him, it's easy to make good money from charisma alone. And since money is society's marker for success, I just accepted it and tried to be like him.

But that's not my style, and for years I was putting on a fake persona. Although it helped me get clients, now, years later, I feel like I let those early clients down. I had no idea, yet, that I was just a salesman taking on way too much responsibility for my clients, something that I wasn't qualified for. But, since it's all legal, and that's how the entire industry works, when things inevitably don't work out, clients just play the blame game and move to another advisor with whom, sooner or later, the same thing will happen to them again. Most clients never truly take responsibility for their own money. I wouldn't be surprised if I get a negative review on Amazon from an old client. They would probably never read this far, but I would say this to them: Sorry, I'm not sorry. I hope you learned something. I sure did. That's why I've done everything I have to write this book and tell the truth. I hope you can take that lesson, too, and go do great things.

My mentor and I just kept churning along together in the same way for years. It worked, and we made money, and we were "successful," if that's how you define success. But I wasn't progressing in the business as far as being a financial adviser working at the firm. I had been working with my mentor for four years, and our relationship was exactly the same as it had been the day that we started. Even worse, the "wake-up call" that I had gotten on the beach in Thailand had started me along the path towards the realization that the financial advice that I was giving people was a lie. My mentor and I were living free lifestyles in Europe, but our clients were all stuck on fixed incomes.

This seemed counterintuitive to me. Why weren't we teaching them what we "really" knew about wealth creation? That's when I lost all faith in my mentor and realized that I had just become a sales tool for him, and by extension, the firm. It broke my heart, and to this day he is still too narcissistic to see it. I realized he was just one of the 99.9% of "financial advisors and planners" who are all giving the same cookie-cutter advice that they also all know is bullshit, since the market and the economy is completely variable and there is never any real guarantee of a favorable outcome. Once I really understood that, I could see that the goals we helped people to plan for were ridiculous. They only caused mediocrity and left everyone feeling disappointed.

I realize now that what I was struggling with was integrity.

My Experience with Apathy

At the beginning of 2014, I finally sold my condo in Germany, and I was ready to move back to the states. That's when my mentor suggested that I move down to live with him in Verona, Italy, instead. He had a huge apartment in the penthouse of this building in downtown Verona, with an office downstairs. He said I could live in the office, and we could get organized and start really working together.

I was psyched, and after traveling for a few weeks, I moved down to Italy that summer. It was what I had been asking for, and I was all in. So I asked him, "What's the plan?" But "the plan" never got any more specific than, "We're going to make some money." He was making four or five times as much as I was already, and what ended up happening was that he was always training for Ironman triathlons, which was fine. I mean, this was the lifestyle he had built for himself. But we wouldn't really get to "work" until we'd gotten up and had an awesome breakfast, because his wife would cook these amazing meals. And then we would go to this awesome gym, and she'd teach a spinning class for an hour. Then we'd work out for another two or three hours. After that, we'd come back and eat this huge lunch at two or three p.m. Only then would we go to the office for an hour or so. What I had assumed we would be doing was that we would get down there and work on the business, follow up with clients, and come up with a business model. Often, though, we wouldn't even get into the office until later in the evening, and then he'd be handling his own clients, and I would just be sitting there clicking on my computer the same way I had with that first CFP. But this time, I didn't waste time trying to learn more financial planning concepts. Instead, I ended up spending a lot of time learning about online marketing, trying to figure out what we should be doing.

What always confused me was that we were focused on the military and veteran community, the federal employees, and the contractors. They are all on these very, very fixed incomes, living these fixed lifestyles over in Europe. Being stationed overseas is nice: you can get a little more time off and travel easily, and you're a little more motivated to do that. You get some housing allowances. In general, everybody's getting a little bit more money than they would have if they were working in the States, and it's tax-free for the contractors. These were the types of clients we were getting.

My mentor and I, on the other hand, were living completely different lifestyles, not only from our clients but from how the average person working in the financial industry lives. We were doing exactly what we wanted to do, every day, while our clients were trapped in unfulfilling lives, and we were being complicit, in a way, in keeping them there, when we knew full well that there were other ways to live.

Maybe in the eighties and nineties there were only a few careers that allowed you to have an independent lifestyle—financial advisers and real estate agents come to mind—but today there are thousands, millions of other ways to make a living in which you aren't tied down to a schedule. What these kinds of jobs all have in common is that their business is built on relationships. You can make a lot more money in relationship-based businesses. Sure, people in these kinds of careers work crazy hours as they get started, but they also have an opportunity to enjoy their life because they are 100% in charge. Since part of their work is building relationships, they can do this while doing whatever it is that they enjoy doing ? golfing or whatever. Independent financial advisers, investment advisers, and even insurance agents fall into this category. I was on the low end of this kind of lifestyle, and my mentor was on the mid-to-high end, and I thought, why aren't we empowering other people to live this way? Why am I not doing what I can in my capacity as a financial advisor to enable people to live this way, if this is what I believe in and what I am doing for myself?

But the disparity between what I was saying and what I was doing went even further than that, like how I wasn't contributing to a Roth IRA. It was mostly out of laziness, but thank God that I hadn't, because if I had had most of my savings in a Roth IRA that would have backfired on me later on when I became an entrepreneur. There is nothing wrong with saving for the future and for retirement, if you want to keep some of that stuff going on. It's another asset class that you can leverage at some point down the road. But retirement accounts are about the far-off future, not the near-future or the present.

Really, though, they are about control. They keep you focused on working in a job and making enough money to hopefully save for retirement so that the government doesn't have to foot your bills. Along the way, the government controls you by taking taxes to pay for supporting all the people who aren't savvy enough to save for their own retirement. The financial tools and strategies that were enabling me to live the lifestyle I was living were not the same as what I was sharing with my clients, and not only could I not figure out a way to explain these tools and strategies to people at that time, I was actually too ashamed to do it! I was ashamed that I was doing something to further my human capital in the way I wanted to experience life. I was ashamed because I was breaking out of a system even as I was entangling my clients up in it.

I wanted to do it a different way. It wasn't that I wasn't identifying with my clients, or that I wanted to tell them all to quit their jobs and go be entrepreneurs. It was that I wanted them to see that a real investment in wealth creation and financial independence didn't mean being frugal and hoping the stock market just works out, and keeps going up, and doesn't crash on the day you retire.

What it comes down to is that I was giving people outdated financial advice. I was still telling them to do what financial advisers were telling people to do in the seventies and eighties, but there's new technology out there. Not only that, technology-enabled financial education is available to everybody. You can Google what you want to know. You can go on YouTube. More and more, people are doing that, and they're starting to understand this stuff. If you're not doing this, you should, because you need to understand your finances. We had clients, though, who never took the time to educate themselves, clients who would let an entire year go by without really thinking about their finances, and then they would want to go over everything with their financial adviser. We would put together a "plan" for them using a calculator tool and print out this hundred-page document and say, "Here's your financial plan." It was very odd. I didn't feel like I was acting with integrity doing things that way. Nobody read through these financial plans, they just threw them on a shelf, figuring they were doing what they were supposed to do. They were saving for retirement. The plan was full of impressive-looking projections, but they were all meaningless, because who knows what's going to happen in the future?

I ended up only spending a little more than four months in Italy. I got tired of feeling guilty about enjoying a lifestyle that my clients would not be able to enjoy until they were in their sixties. We never generated any new business while I was down there. Nor did we try to come up with a business plan, so that was frustrating. My mentor was busy training for triathlons, and he would take me on these beautiful, awesome bike rides. It was a great experience. But after a few months of this, I had figured out that he got bored, and I was just kind of his workout buddy. I decided I wanted to go back to the states to spend Thanksgiving with my family. He seemed a bit insulted. I didn't even get a ride to the airport or a goodbye.

My Journey Into Aspiration

By January, I was settled in San Diego. By this time, I had some good money in the bank, but I was very unhappy, and thank God it all happened that way, because when I left, I knew that now I would be able to find my purpose, my mission. At one point, I had thought that I couldn't stay in the financial industry and still have integrity, but eventually I realized, oh my gosh, I can actually offer people hope for change. I can explain things differently. I can talk about what role my clients are in and what identity they want to have. I could define my own identity as what used to be called a financial planner. I was ready to define what that meant to me and for me. I wanted to talk to people about how to set themselves up for how they wanted to live their lives.

So when I got back to the States, I didn't have any motivation to continue doing what I had been doing the way I had been doing it. I did a couple of sponsorships, but I hardly followed up with anybody. Things fizzled out with my mentor pretty quickly, and I felt kind of abandoned. And even though I had had my epiphany, I just kind of moped around for a few years when I should have been saying, "Ah, this is it!" and taking some kind of action towards merging my actual identity with my business. But though I hired a web designer and started going to conferences, I was still trying to be the traditional financial adviser, with a focus on military and veterans, because that was the life situation I was most familiar with. But at the end of the day, I was still giving them the same type of advice that I had been giving and which I no longer believed in or felt good about giving.

At the same time, I was also going to seminars, and learning how to write a book, and learning more about how to start a thought-leadership-type business. I was becoming aware that this kind of business was out there, that there was a sliver of the thought leadership industry across pretty much all business disciplines, and it was amazing. I was starting to see how these influencers and these businesses work and what they teach people. And I wasn't just passively consuming information; I was trying to peek behind the curtain. It was from the financial thought leadership business that I learned to redefine what real financial guidance is. It was obvious to me from the get-go that acting in my clients' financial interests meant having conversations with them that went way beyond retirement planning. It meant talking to them about protecting their lifestyle, helping them to create something they really wanted to do, protecting their wealth and their influence, helping them to control the source of their income, and leveraging their asset classes, which included their own human capital and knowledge and how they looked and networked for opportunities.

Achieving Autonomy

When I first got involved in the financial advisory business, in 2008, I thought, "It's only money." But I came to realize that people who want to create, people who are thinking differently from the herd, don't need a financial advisor. What they need is a guide, someone who can point them in the right direction and give them the tools they need to figure it out. These kinds of people need more than retirement planning in the traditional sense. That doesn't mean that they don't want to have 401(k)s and IRAs, it just means that's not *all* they need to have. Their financial planning needs to include different ways of saving money that allow them to actually *access* that money in the *present* if they want to try out an idea or make some change in how they are living. And so often, this kind of individual was just keeping their money in a checking account, which isn't efficient.

And so I thought, why don't I take all of these things that I'm learning and implementing to create my own wealth and liberty and start providing *actual* financial advice that allows people to create opportunity for themselves? I knew that the way that I understood finances, and the way that I wanted to work with people, and the kind of people that I wanted to work with, meant that I needed to find a way to connect with the kinds of clients who stood to benefit the most from the kinds of services I wanted to offer. I believe that the more honest you are, the more open you are, the more you share with people who you are and what you're learning from things and why you did or didn't do certain things, that's going to resonate with those whom you're meant to serve. Being transparent helps to weed out the people you aren't meant to serve and reveals the people that you want to be working with.

For instance, the reason that I talk about LinkedIn so much on my website, and the reason I am going to talk about it at length in this book, is because I believe that of all the money tools like stocks, bonds, mutual funds, banks, and real estate aside, the best wealth management and wealth creation tool right now is free, and it's called LinkedIn! You can create wealth on that platform, if you want. It may require you to invest many years of your time, but it can happen. It reminds me of the quote that goes something like: most people will spend 40 years doing something they hate to reach retirement, but won't sacrifice a paycheck and put the work in for four years to build the knowledge and skillset they need to live a life they love. Well, I've taken almost exactly four years without a paycheck, and when it comes to skillset development, LinkedIn was the thing that put me over the top. I've been on the platform for 12 years as of the writing of this book, and in the last 18 months I have arrived at a whole new level. In two different instances, I took a course in a single day, learned a bunch of stuff and started implementing it, and things really changed. It was like getting a high. I did it again, and I realized, oh my gosh, anybody can do this, and they should. It's better for everybody if we all learn how to use LinkedIn better. I am so passionate about this topic because one of the ways to create more liberty for yourself is to control the source of your income.

It took me five years of investing in training, spending money starting a few different brands and businesses, and failing at them before I finally figured out what my brand identity is. Having that understanding of who I am and what I am about, what I'm going to do about it in the world, and how I'm going to serve others in that capacity has nothing to do with money, and yet it's the very foundation of wealth. Let me repeat that, because it's super important:

Understanding who you are and what you're about, what you're going to do about it in the world, and how you are going to serve others in that capacity is the very foundation of wealth.

If I was able to make six figures as a financial adviser when I didn't know anything, then the sky's the limit for me now that I am able to differentiate myself from other people in the industry with my own unique value proposition. Not everyone wants to be on the traditional financial path, and for those that don't, I can show them that there actually IS another path.

In the first years that I worked as a financial adviser, my clients didn't have anything in particular in common. I served young adults and grandparents, officers and enlisted. There was no single message that I could offer them and no single way that I could make a difference and do something meaningful. But once I realized my brand identity, I wanted to get real and see if I could find ten other people who wanted something more than a salaried position in the rat race; I wanted to see if I could find ten people who were willing to take risks in order to stay *out* of the rat race. And I found those ten people, and then another ten people, and then another... Today, all of my clients have this thing in common: I serve military and veterans who want to create and seize opportunity.

It's staggering, how amazing it can be to stick with entrepreneurship, to be willing to take risks because you believe in something. When you're creative, you're willing to try things.

You will fail before you succeed, but if you keep at it, the only way you won't succeed is if you quit.

The Big Takeaway From My Personal Story

I titled this chapter "The Modern Economy," yet I hope that you have noticed that I haven't talked that much about technology. That's because the modern economy is not about technology. Technology means the application of knowledge for practical purposes. As I said in the beginning of this chapter, technology is only a tool. It's an enabler. The modern economy is about identity. It's about how technology is enabling more and more people to add value to the world by bringing their inner and outer lives into alignment which, oh by the way, also happens to make people very happy.

People today are earning money in the modern economy because of *who they are*. They are sharing their personal stories, their interests, their knowledge, and their experiences with others who are looking for exactly what these individuals have to offer. The products and services that are available are as varied and unique as the individuals who offer them.

If you want to participate in the modern economy, if you are currently stuck in apathy, if you know you are not happy and you want to be happy, I can show you how to get there. All you need to bring to the table is your willingness to grow. This means that it's okay to not know exactly who you are or what you want yet, but it's not okay to not be seeking it. This book is not about your career. It's not about your retirement plan. It's not about your financial plan or your salary. This book, like wealth itself, is about what all those things are for. We all need money. We should all save money. But if you never try to align money to your purpose, then you're just one of the masses who is destined to live and die in a state of apathy. If you want more than that out of life, then turn the page, because we're about to head into aspiration.

Part II

Into Aspiration

6

The Power of Why

This is the real secret of life—to be completely engaged with what you are doing in the here and now. And instead of calling it work, realize it is play.
~ Alan Watts

In the last chapter, I shared a big secret with you:

Understanding who you are and what you're about, what you're going to do about it in the world, and how you are going to serve others in that capacity, is the very foundation of wealth.

Aspiration is the name I have given to the journey of getting to this point of understanding and taking steps to do something about it. We'll get into action steps in a later chapter. But before we do that, I want to spend a little time talking about "understanding who you are and what you are about," because for most of us, getting here is no small feat.

Maybe you have always had this understanding. Maybe you were born with it. Some people have just "always known" who they are, who they are destined to be, and how they are best suited to add value to the world in that capacity. That doesn't mean they aren't stuck in apathy. Knowing what you are meant to be and do is one thing, taking action to be and do it is something else entirely. When it comes to pursuing our passions, it's very easy to get derailed. We can get derailed by well-meaning parents who don't really understand us, by where and how we grow up, by the kind of education that we get, and even by random events in our personal history. Our own limiting beliefs can derail us, too, as I'll talk about in a later chapter. Responsibility, with a capital R, can also thwart us if we think that doing what we want to do and what makes us happy is somehow inherently irresponsible. We all ultimately have to take responsibility for ourselves. We have to find some way to keep a roof over our heads and food on our tables, and we don't always grow up realizing how many different ways there are to do that. Responsibility is the trump card of the status quo. In all too many cases, giving up on our "impractical" dreams is simply one more right of passage on the way to becoming an adult.

If you're like my friend Pete I told you about earlier, who constantly told me while we served together that he wanted to be a physical therapist, and you already know now who and what you want to be, then you are one step ahead of the rest of us. Pull your dream out of the back of your closet, dust it off, and prop it back up on its feet. You're going to need it. But if you don't already know what your purpose is in this world, then you are going to need to buckle down and do some "work." In either case, before going forward, you are going to need to make sure you know your why.

Knowing your why is the foundation of happiness, and happiness is the foundation of wealth. I am going to take a moment here to talk about those two things, happiness and wealth. They are both integral to the concept of autonomy. And while both have financial components, it is important to understand that neither of them, in and of themselves, are about money.

What Happiness Is—and Isn't

Are you happy? Have you ever even asked yourself that? Or, even worse, maybe the answer is so obviously no that it makes the question itself unnecessary. It's a dangerous question, because sometimes if you are not happy, getting to happy requires a lot of disruption, and we tend to fear and dislike disruption. Leaving unfulfilling relationships or jettisoning "successful" careers that make us miserable are massive steps that can often feel more like setbacks than progress, at first. But those are worst case scenarios that can often be avoided simply by asking ourselves this question much sooner in life.

Ask yourself now: Are you happy?

Happiness is not just an emotion like joy or contentment. Neither is it just a mood, like cheerfulness or satisfaction. Many emotions and moods can express happiness, but the intermittent presence of any particular positive emotion or mood does not necessarily indicate that a person is fundamentally happy. Emotions and moods come and go. Happiness is a pervasive, underlying state of mind that includes an understanding of your why. Happiness comes along with having, pursuing, and expressing a sense of meaning in your life.

In a sense, happiness is independent of the people and things around you. You can be happy and still experience negative emotions and moods. Happiness is about being comfortable in your own skin. It's about looking forward to getting up every morning and living your life. You can be excited about some event or belonging; you can be satisfied with your home, your family, or your ability to pay your bills. Feelings of excitement and satisfaction can occur even if you are fundamentally unhappy. In fact, they frequently come along with the pursuit and satisfaction of temporary distractions from unhappiness. But those feelings aren't themselves happiness because they are attached to something in the physical world and those things can be lost. Also, things which make us "happy" in the pursuit and satisfaction of distraction often don't stimulate that feeling for very long. Once the novelty wears off, we tend to look for the next thing we want, the next temporary high.

If what you interpret as your happiness would disappear along with its physical markers or manifestations, then what you feel is not "happiness." Happiness cannot be taken from you. Happiness transcends the inevitable dips and setbacks in life and recognizes them as the (perhaps temporarily disappointing or painful) emergence of some new opportunity. Happiness is part of the essence of your being. It continually manifests in the physical world, too, but even if its markers disappear, the essence remains and new markers can be created. That power of creation and potential comes along with happiness and cannot be taken from you either. These things are aspects of your very being and everyone has them, whether you know it yet or not, and whether or not you feel capable of accessing and expressing them right now.

The presence of happiness is the difference between someone who is pursuing a passion and someone who is just following the status quo, even if those two individuals are practicing the exact same career in exactly the same way. The status quo is the shape or shell of something without a why at the core. Where there is no why, there is no happiness. At best, there is merely one of its many counterfeits.

Sometimes it takes an experience of massive unhappiness that feels beyond our control for us to begin thinking deeply about who we are, what we're about, and what we are going to do in the world because of it. Remember the German opera singer I mentioned in the last chapter when I told you my story? It was the breakup of that relationship that got me started asking these kinds of questions. The loss of the future that I thought I was going to have and that I thought (or wanted to think) was right for me was my first serious experience with unhappiness. I had to sit down and think about what I wanted from life, not just from another person, but from myself. I was thirty-two years old, and I had never thought that way before or asked myself these kinds of questions. Up until that point, I had simply been carried along by the structure and expectations of my family, school, college, and the military. I had no idea how to approach questions like what do I want, am I happy, or how can I become happy? It felt weird and uncomfortable even to have these thoughts. Now it seems weird, uncomfortable, and downright tragic *not* to.

Happiness and Money

Like many people, I have had my share of times in which I had next to nothing. I have also had my share of times in which I had plenty, much more than I needed to meet my basic needs. As I was going through the identity crisis and career evolution that I described in the last chapter, I discovered that happiness has little to do with money. I found this out for myself, when I reached the point of having plenty of money and found that it didn't change the way I felt about myself. Making money in and of itself was not fulfilling to me as a purpose or a goal. But I am not the first person to find that the correlation between money and happiness is not as large or as consistent as we might think.

Numerous studies have looked into the relationship between money and happiness. What they have in common is the conclusion that more money only makes you happier to a point. Different studies have tried to put specific dollar figures on that point. One says it's $75K, one says it's $105K. But the real point here, I think, which gets hidden when you start talking about dollar figures and whether you live in a rich country or a poor one, is that money has the greatest impact on happiness with regards to whether or not you can meet your basic needs: food, clothing, shelter, hygiene, health. How much it takes to meet those basic needs depends on different factors. For example, the cost of living in some geographical areas is significantly higher than it is in others. Even though two individuals may live comparable lifestyles in every other way, that lifestyle can have vastly differing price tags.

The amount of money it takes to meet your basic needs also depends on your responsibilities. A person who cannot meet their responsibilities will naturally feel stress, and prolonged stress can lead to unhappiness. A single person and a person who is married with six children have a different level of financial responsibility and different needs with regards to how much money is required to meet basic needs. The actual dollar amount, here, is relative.

And then there is the other thing that factors into happiness, which seems to exist independently of any dollar amount: whether or not a person is doing work that they want to be doing, that aligns with their sense of purpose and life mission. Many people want more money so that they can have more time and freedom to do the things that they enjoy. What's left unsaid here is that what they spend most of their time doing ? their job ? is NOT something that they enjoy. People who are happy with the work they are doing do not count the days until their next paycheck or get depressed on Sunday night because they know they have to go back to work the next day. When you are doing work that makes you happy, the work itself rewards you, and the paycheck becomes less significant. It matters less what day of the week it is; you work when you want to work, and you want to work, because work feels like play. You feel free instead of stuck; liberated instead of frustrated; excited instead of bored.

I know people who meet their basic needs living in countries like Thailand where the cost of living is low and no one is trying to keep up with the Joneses. They have a yoga practice or work in the travel industry, for example. That's all they need. They live a very nice life and are extremely happy because they got out of the 9-to-5. Similarly, I know people who are very happy earning hundreds of thousands of dollars a year. They aren't happy because of their income; they are happy because they are doing work that they love which happens to generate a substantial income.

The takeaway: Money is a component of happiness with regards to the ability to meet one's basic needs and responsibilities without undue or prolonged stress, but the further you get beyond that point, the less money matters. The amount of money it takes to comfortably meet one's basic needs and responsibilities differs from person to person and situation to situation.

Figuring Out Your Why

In *Life, Liberty, and Purpose,* I talked about the three key things that are important to understand about money:

- Money is about confidence.
- Wealth isn't about money?it's about time.
- Money is not your why?Money is the means to your why.

I have already alluded to the first two bullets, although not in so many words. If you want to dig into them, you can check out the other book. But I do want to look briefly here again at the third one.

Money is not a why.

To figure out your why, you need to figure out what work you would do in the world whether you got paid to do it or not. When answering that question, don't get distracted by things like whether or not you think what you like to do is a "viable business option" or whether or not you think anyone would ever pay you to do it, or whether or not you can conceive of a career path or business right now that has anything to do with what you love to do. Just answer the question. Put it out there.

What work would you do in the world whether you got paid to do it or not?

Here's a hint: you are probably already doing this work and not getting paid for it. Look towards what you do in your leisure time. Look towards what makes you feel good. Look towards the things for which people seek out your advice or expertise. Look towards the things that you know how to do and that you enjoy continually learning to do better. That's probably where your answer lies.

Another approach is to think about what kind of person you are. This is a slightly different way of asking yourself what your life's work is or should be. Some people are doers and thrive on physically doing something. They're all about, "Give me tasks; let me do something with my hands; I'll knock out those projects, and I'll help you out that way." Some people are leaders and thrive on strategizing solutions for a group. They're all about, "I'm here to help you achieve that goal or organize in case of an emergency situation." Some people are teachers and thrive on sharing knowledge so that others grow and become more of who they are. They're all about, "Here is how to find that out; here are three new ways to do this. Try it out, and I'll be here if you need me." Some people are creators. They feel the most fulfilled when they are putting something new into the world. They're all about, "I want to figure out and share this new perspective, process, product, or methodology," or "I want to put my own twist on an old idea or interpret something that is already out there in an entirely different way."

Still not getting there? Here's a third tactic, the one that ultimately made all the pieces fall into place for me: Identify who you aspire *not* to be.

Who I Aspired *Not* to Be

I did not grow up "just knowing" that I wanted to help veterans gain personal autonomy through financial control. Before I figured out my own aspirations, I learned who I aspired *not* to be.

In the last chapter, I told you about how I kind of fell into the financial industry when I got out of the military and floundered there for awhile. I told you about my epiphany on the beach in Thailand and how my identity crisis ultimately led me to the realization that I could not in good conscience continue to give people financial guidance that funneled them into and kept them trapped in the status quo when I knew better and managed my own life quite differently. That was a high-level version of the story. As it unfolded, though, there were other additional epiphanies that pushed me closer to ultimately finding my own why, and I am going to tell you about them now.

I told you in an earlier chapter about how I saw a top advisor in our firm give a presentation about his "holistic financial planning" methodology. As I said, I was quite taken with this presentation, which appealed to my appreciation of expressing things simply and concisely with graphics. It also appealed to me because I enjoy getting up and speaking. Unlike boring PowerPoint slides full of words that people sometimes stand up and read verbatim, his presentation was visually compelling. It was meant to be presented.

Meeting this person was a turning point for me. I already told you how we entered into an agreement to work together, and how that change of direction resulted in me beginning to earn some decent money. But that was only the first of two turning points I experienced during my relationship with this person. The rest of the story is that as I worked more closely with him, I came to see that his "holistic financial planning" was not a methodology at all, it was just a sales tactic. We used this presentation on clients for years, but the planning itself was all done using a spreadsheet, and was no different from what anyone else was doing. Whenever a client would get upset about anything, he'd just scoff at their ignorance. My second turning point with him was the day that I realized: Holy shit, he doesn't care! This is all a facade. He's just so proud of himself and the fact that he's the top guy in the firm that everyone looks up to.

In five years of working with him, I hadn't grown in my career at all. I was no different than I was the first day that I met him when he took me under his wing. I was not a partner. I was just a lead generation tool for him.

My epiphany: I did *not* want to be just a salesman of financial products who ultimately did not care what effect the products I was selling had on the lives of the people who were buying them.

I also told you that during the time that we "worked together," I spent a lot of time clicking on my computer, learning about online marketing, trying to figure out what we should be doing to create a unique brand. I had also started reading books about making money while I was still in the military, and continued to do so afterwards as I worked in the financial industry. One of these books was *Rich Dad, Poor Dad*. I learned about what the author of that book, Robert Kiyosaki, calls the cash flow quadrant. You can look this up online. I also highly recommend that you read the book. The basic point of the cash flow quadrant is that everyone falls somewhere into a quadrant consisting of four money-related roles, two in each quadrant. You can draw this for yourself as we go along:

On the left side of the quadrant are the employees (E) and the self-employed (S). They are focused on security. The holy grail for the employee is a good job. When the employee wants to earn more money, they look for a higher-paying job. The self-employed have a higher tolerance for risk, and are usually engaged in service-related careers. While they may consider themselves entrepreneurs, because people in this category only earn money while they are working, in the words of Kiyosaki, "they don't own a business, they own a job." When the self-employed want to earn more money, they raise their rates and/or put in more time. People on the left side of the quadrant trade their time for money. They also pay the bulk of the taxes in this country.

On the right side of the quadrant are the business owners (B) and the investors (I). A business owner owns something which makes money for them whether they are hands-on or not. Business owners hire people that have the skills their business needs. A business owner sees employment as the risky option. While an employee is subject to forces beyond their control, a business owner is in control of his or her own destiny. Investors have an in-depth understanding of finances. The investor identifies and puts money into assets that create cash flows for themselves. They then use that cash flow to get more assets. An investor has a passive income, meaning they earn money off the work that other people do. Business owners and investors do not trade their time for money. They trade their money to create or invest in assets that allow them to use their time as they see fit. They also pay the least in taxes.

When you work in the financial industry, you are not an employee in the traditional sense. Because you only earn money when you have your own clients, it is closer to self-employment. In fact, some firms will even tell you that you are a business owner. I believed this about myself, until I learned about Kiyosaki's cash flow quadrant and realized that I was really still firmly on the left side.

I already knew that I was selling my clients into a status quo existence that I, myself, didn't want to live, didn't believe in, and knew to a large extent how to avoid. My new epiphany was that as long as I remained on the left side of Kiyosaki's quadrant, I wasn't really as different from my clients as I had imagined myself to be. If I didn't want to be a part of the status quo, I needed to aspire to more than just earning and holding onto my money. I needed to aspire to create a business.

The question for me then became, if I want to continue working in the financial space as a business owner, how do I differentiate myself and my services from what everyone else is doing? How do I add value in this space?

You Have to Figure Out Your Why

As you can see from my story, figuring out your why is a step-by-step process of exploration and course correction. Do not expect it to happen overnight. Devote some time to this. Observe yourself.

You have to go into figuring it out. Not just what you want to do, or how you are going to do it, but *why* you *have* to do it. If you think you already know, go deeper. If you're saying, "I want to open a business after the military," "I want to start a nonprofit," "I want to go back to school..." that's all the what. That's not the why.

The what is an expression of the why.

I will say it again: The status quo is the shape or shell of something without a why at the core. Where there is no why, there is no happiness. This is true no matter which side of the cash flow quadrant you are in.

Say you want to do something. But why do you want that? You need the complete answer to the question, "Why? Why are you here?" If the answer to this question doesn't interest you, or if you aren't comfortable even asking the question, then I'm surprised you have already read this far. Asking yourself why, understanding the why of your own life, is the foundation upon which everything that I have to teach you depends.

My own answers to my why began with the realization that I couldn't, with integrity, just follow the masses that are doing the same old things over and over again. I am here on this earth to wake others up, to help them see what's going on around us. I can't honestly be the financial advisor that just tells people to save for retirement. I can't honestly say, "Oh, just get out of the military and get a good job, Sergeant, you'll be happy once you get that six-figure income." It's a lie. It's not being true to who I am or what I can do to help others. I want to guide people how to use the benefits and opportunities for wealth that are right there in front of them, because there's no better time in history to be a veteran in the United States of America. I'm here to help veterans live with a purpose. I'm here to help *you* live with a purpose. I'm here to help you develop your happiness in life around that purpose. Even if I only help one person, then I've accomplished something valuable to me.

One More Secret...

When you are *not* acting according to who you are and what you are about, life can feel like a struggle. As soon as you do begin to act according to these things, clarity will open up for you about how you will create the life you were meant to live. When you know who you are, who you connect with, and how you like to serve, everything will fall into place.

This doesn't mean that there won't be problems and setbacks. But you will not take them to heart or let them stop you because your perspective will be different. You will view them as opportunities rather than roadblocks. As you meet and work through each challenge, you will continue to move forward and inevitably come out the other side better off and closer to where you want to be than you were before. Struggles and setbacks are merely the birth pangs of opportunity. Embrace them. They are your friends.

There are a lot of books and videos out there about this concept and about the "physics" of the human mind, and I am not going to get into that here. But trust me when I tell you, *this is true*. If you don't believe me, then simply start taking steps in a direction that you know, deep down inside with every fiber of your being, that you want to go, and watch what happens. Things may not happen in the way that you expect. In fact, you shouldn't have any expectations about *how* things will happen at all. But if you learn the skill of recognizing and pursuing opportunity as it arises, whether you feel prepared or ready for it or not, you will see for yourself that the life of your dreams is just waiting to unfold before you.

7

Be Honest and Become

Your best is going to change from moment to moment; it will be different when you are healthy as opposed to sick. Under any circumstance, simply do your best, and you will avoid self-judgment, self-abuse and regret.
~ Don Miguel Ruiz

I spent the whole last chapter talking about how important it is to figure out your passion and your purpose. In a sense, this is all about honesty. You probably don't think that trading your time for money at a job that you don't really care about is dishonest. But showing up in the world as something other than what you are *is* dishonest. It eats away at you over time and leads to unhappiness. Have you ever felt insecure or a little paranoid at work (or even in a relationship), afraid you were going to be "found out" because you really didn't care, didn't want to be there, or didn't think that you belonged? Have you ever looked around you and wondered if you're the only one who is "acting" instead of living?

This book is about being honest with yourself and with the rest of the world. It's about bringing who you are on the inside into alignment with who you are on the outside. It's about discovering the best way to express *you*. Regardless of your individual why, you are here to add value to the world in some unique and personal way. You are part of a larger puzzle, and if you are not living an authentic life, then you aren't just cheating yourself. You're cheating everyone in the world you are meant to serve.

But there is another aspect of being honest. In order to realize your potential, it is not enough to discover and understand your own unique value. You must also discover and understand—and get rid of—the things that are holding you back.

The Invisible Enemy Within

We all have things that hold us back, even though we don't always realize what they are. Or rather, we know that we think certain things about ourselves or about the world, but we don't realize that our thoughts aren't always necessarily true. They are only true to us.

We can change what it is true to us, and when we do, we change our lives and change the world that we live in.

Much of what we think and believe about ourselves and about the world has been programmed into us by circumstances and the people around us. When we make decisions and go about our lives, we think that we are thinking, but all too often we aren't thinking at all. We're allowing these background programs to run our lives on autopilot. These programs can be crazy powerful. They are so powerful that we feel "good" when they are in charge. But "good" in this context doesn't necessarily mean good. It means that they keep us feeling the way that we are used to feeling, even if what we are used to feeling is angry or frustrated, defeated or unhappy. Sure, we may spend a lot of time thinking that we want things to be different, but we are so used to these negative emotional states and the chemicals that they manufacture inside of our bodies, that when we do something different that makes us feel something different—like happiness or joy—it can feel...weird...and extremely uncomfortable. This discomfort can manifest as anxiety, fear, and/or negative self-talk. Maybe we get a little warning thought, "Hey, we don't usually feel this way...is something wrong?" And we hurry back to the way we are used to being, feeling, and thinking.

These ultra-powerful programs are called limiting beliefs, and one of the ways that you get beyond apathy is to find these programs that are always running in the background in your own brain, see them for what they are, and shut them down. How do you do that? Watch yourself as you go about your life. Listen to what you say to yourself. Then, decide what you want to do and do it, consciously, instead of allowing a limiting belief to make a decision and direct your actions for you. That, and tough out the temporary feelings of discomfort that will inevitably come along with choosing to act in a new way.

You must discover and dismantle your limiting beliefs. They may have convinced you that they are your friends, but they are responsible for keeping you in a state of apathy.

Limiting Beliefs Explained

A limiting belief is something that we tell ourselves about ourselves or about the world and how it works. Limiting beliefs shape and limit how we live our lives. We all have them. These insidious little programs often come from our parents or other influential people in our lives, who probably had no idea that they were even installing them or how damaging they could be. Here's a few examples:

- You're stupid.
- You can't do that.
- You aren't attractive.
- You're a burden to me.
- You'll never amount to anything.
- You're just like your father (or mother).
- You should be more like your brother (or sister).

Can you feel how just reading this string of horrible statements pulls negative energy towards you?

As we get older, we have more and more experiences. We respond to the outcomes of those experiences with either positive or negative (or maybe even neutral) emotions. We assign thoughts to those emotions which make sense out of them and connect them to the event that aroused them in a way that seems logical to us, given the information we have available. It is the job of our thoughts to make sense out of the world around us, and they will go to just about any length in order to do that, however crazy these thoughts (which we call "reasons" or "explanations") may seem to other people who have lived different lives and had other experiences.

The negative thoughts that we assign to unique and individual events which aroused negative emotions can become limiting beliefs, too—automatic programs that we install ourselves and which gain more and more power as we repeat them to ourselves over and over again. Here are a few examples:

- No one wants to hear what I have to say.
- Other people are stupid (cruel, better than me, etc.).
- I can't...
- It's stupid and irresponsible to...
- If I....bad things will happen.
- X, Y, or Z is always the "wrong" thing to do.
- It's risky to...
- It's risky *not* to....
- "Good" mothers (or fathers, sons, daughters, wives, husbands) don't...
- I don't know how to do anything else.
- There's nothing else out there for me.
- No one will ever love me.
- Nothing that I do makes any difference.
- There's no way to fix this.
- Things will never get any better.

I could go on, and on, and on. You probably could too. The problem with limiting beliefs, and the reason that you *must* discover what yours are and stop feeding them with energy, is that they inevitably manifest themselves in your life. Henry Ford said it best: Whether you think you can or think you can't, you're right.

If you think that you will fail, you will fail.

If you think that it's not possible, it isn't.

If you think that what you have in your life right now is all that you can ever have, it is.

The good news is that this same mechanism works just as well in reverse. Remember in the last chapter when I told you that as soon as you begin to move in a direction that resonates with who you really are, that everything will fall into place? The reason for this is that you will be thinking in a positive way about what you want. You will hold your vision in mind not just as a possibility or an idea but as a belief. Once you do, you will inevitably draw that future towards you.

The choice is yours.

I said above that these programs tend to run us without us realizing it. However, once you start looking for them, they are surprisingly easy to find, because they tend to be at the bottom of everything we do, every decision that we make, and everything we tell ourselves about what is going on in our lives and why.

The key is to look.

Limiting Beliefs in Action

In the last chapter, I told you about a few of the additional mini-epiphanies I have had on my own journey to autonomy. Here's another one.

This story takes place when I was working as a financial advisor in Germany, and it's about a young woman that I always considered my favorite client and one of my best friends. She was a former Air Force officer, really nice with a super-jolly-go-lucky type of personality, who was working over in Germany as a Project Manager for one of the DoD contractors overseas. Those jobs tend to suck the life out of people. I had many friends that were just miserable with the hours they were putting in and the stress that was on them. Government contracting is high stress, because if you lose the contract, the company doesn't get any money. People are constantly getting laid off when contracts end.

This woman was obviously very productive and didn't really have much risk of getting laid off. She and many others like her justified doing this painful work because, guess what? Government contracting pays pretty well. In fact, they're easily making the equivalent of $150,000 dollars a year, and the majority of it is tax-free due to the overseas foreign income exclusion tax. Of course, when you're working overseas you're also getting extra housing allowances. As a single, 38-year-old former Air Force officer, she was earning a high upper-middle-class lifestyle. She was also frugal. She wasn't cheap by any means, she just didn't need her money. She lived in a small place and saved a bunch of money on housing. She liked to go on weekend trips and really enjoy life. She took advantage of any moment she could get away from work to travel. Her Facebook photos showed it. It was amazing to watch her. Eventually, she started to fall in love with going down to Africa to work with nonprofits and charitable organizations in Kenya and Ethiopia.

Now, one thing I haven't talked about yet was the fact that because of this high income and frugal lifestyle, she had managed to save $750,000—the majority of it in retirement accounts—by the time she was 38! That is not something you see very often in the military and veteran community (or anywhere else, for that matter).

It's really not that hard, but she was an anomaly at the time. I have seen Senior officers, O6 and colonels, coming out of the Army broke and in debt, not having saved any money, and scared to death of what comes next. So I'm sure she thought she had a stable lifestyle. But every time we'd get together, she was always confused about how we did the retirement planning. She didn't want to take risk in the markets. So we had her invest in low-risk vehicles, vehicles that offer absolute guarantees, in case the stock market crashed; vehicles that had no cost involved in them, and then some that had other costs. Either way, it was a completely unique and diversified portfolio because she was a very unique person and had unique opportunities around her. She didn't need to take the risks in the market that her colleagues were taking. (They were bragging about their returns, back then, but I'm curious about how their accounts responded to the COVID crash…but I digress).

Now, I was never in a position as a financial advisor to give any life advice. I was just supposed to manage the money. Again, I was doing it based on the old school methodologies, and what it came down to is I recommended what my mentors told me I should sell her.

Then she got very passionate about these trips she was taking to Africa. She started to get more engaged with how her accounts were doing. Two things happened. The first was that she started to question how we were investing her money. I would remind her that she didn't need to take risk and that she had said she didn't *want* to take risk, and I would re-explain what the fees were for. There was probably someone talking in her ear because his TSP was going up higher, which had nothing to do with her. But what are you going to do when someone's buying Apple stocks and getting returns, and boasting about their big gains, all the while not understanding the bigger picture of how the market works?

The second thing was that she asked me, "What if I used $20K and went down to Africa and travelled the world for a year, quit my job. Could I do that?" She was so nervous about the idea of quitting her job and using her money. This is about the time when I started to have problems with financial planning. So I said, "Yes! Take the money. Use $50K. You have enough. You don't spend it. Use your money. Go enjoy the life you want."

And then she fired me!

She sent emails to my mentor saying that I was in over my head. That's when I realized that I wasn't in over my head — everybody else was doing financial planning wrong. She had literally created the opportunity to do whatever she wanted, but somehow she had it in her mind based on some limiting belief(s) that she couldn't do it. That's when I knew that if I was going to be a financial advisor, advising people on their portfolios was a waste of time. Anybody can Google how to do that. If I'm going to be a "fiduciary" and make sure I'm doing things in the clients' best interests, her best interest was to use her damn money to go enjoy life and do the things that filled her up.

But unfortunately, I didn't know how to communicate that, and this inability to express what I understood and believed about money set me off. I needed to fix this. I decided that if I was going to do what's in the "client's best interest," I was going to teach them how to use their money to create wealth and abundance and opportunity, because there's never been a better time in the world to do that than now. Unfortunately, my old friend is still working overseas as a government contractor, miserable. I hope she knows that she inspired me to help tremendous amounts of veterans change their lives and use their opportunity. I hope one day she joins me.

My epiphany: Limiting beliefs limit your life.

Self-Awareness = Success

Limiting beliefs can hold us back, but that doesn't mean that we should believe we can do anything and everything ourselves or that we should even try to. We all have things we do well and things we don't do so well or which don't come very naturally to us. This is simply a fact of life. At some point in your journey of personal growth, there will come a time when you need to determine what it is worth your time to do, learn to do, or get better at, and where you should apply resources instead because your personal time would be more happily and/or more productively spent elsewhere.

You may already be aware of some of your weaknesses. It's hard to get through life without realizing them. When they reveal themselves, it's usually because they have just knocked us on our ass in some way. Unpleasant experiences aren't easy to forget, and we tend to form or reinforce limiting beliefs around them. As you proceed down your new path of aspiration, you will be learning, and your world will be expanding. As a result, you will probably discover new weaknesses. While you should not let these stop you, don't ignore them either. Learn to tell the difference between a limiting belief and a real weakness. Some things that we initially perceive as weaknesses aren't weaknesses at all. They're just the learning curves we have to face as we approach and begin to master new skills. You can learn to do some things within a reasonable amount of time. If it seems worth it, and you want to learn this new skill, then by all means do that. The more you can do, the more powerful you will feel. You may even find that you have strengths you never knew you had.

However, there's no reason to invest time learning how to do every single thing you might want or need to do by yourself, and I don't think that you should. There are some things that we aren't good at. They don't come naturally to us. We'll never be able to do them really well, and frankly, we don't want to do them anyway. These are the places along your journey where I encourage you to apply resources to accomplish things instead. For example, when I told you my story, I mentioned that my mentor would have me sitting and taking notes while he conducted meetings with potential clients. I don't like taking notes, and I suck at it. At the time, I didn't realize that scenario could have unfolded in any other way. I did not understand then, as I do now, where to apply myself so as to get the most ultimate value out of my time. That is a great example of what I am talking about. I could have hired someone to sit there and take the notes for me, so that I could have been more engaged in the meeting and learned more, more quickly. Even better, I could have recorded the conversation and had the audio transcribed, and then paid someone to edit the notes into an executive summary. I probably could have done all that for $100. Most financial advisors would scoff at the idea, but given the amount of money you can make for a client, this is a reasonable expense if it adds value to your business. The sooner that you become aware of these areas of weakness in your life, the more quickly you will be able to deal with them in a way that turns them into areas of strength, not for you personally, but for whatever enterprise you decide to pursue.

One of the amazing and empowering things about the modern economy is that it is full of talented people who already know how to do the things that you can't or don't want or don't like to do. They can do them far better than you will ever be able to and in much less time. This pool of talent makes up what is called the "gig economy," a virtual support system of virtual assistants, transcription services, content writers, ghostwriters, editors, artists, videographers, graphic designers, web designers, SEO specialists, social media specialists, art directors, etc. who are available to help you and who are probably pursuing their own passion in doing so. Leverage this pool of talent. It is your friend.

Knowing your genuine weaknesses and what to do about them is just as critical to your eventual and overall success as knowing your limiting beliefs, your passion, your purpose, and your strengths. Everything I have talked about in this book up till now all comes down to one thing:

Self-awareness is the key to success.

In fact, a fascinating 2010 study that looked at 72 executives at public and private companies with revenues from $50 million to $5 billion found that a key trait common to all of them was self-awareness: "Interestingly, a high self-awareness score was the strongest predictor of overall success. This is not altogether surprising as executives who are aware of their weaknesses are often better able to hire subordinates who perform well in categories in which the leader lacks acumen. These leaders are also more able to entertain the idea that someone on their team may have an idea that is even better than their own."[1]

This need for self awareness is equally important when it comes to the "soft" skills. The same study noted: "The qualities commonly associated with management and leadership, being authoritative, decisive, forceful, perhaps somewhat controlling, if not moderated by a high degree of awareness as to how one comes across and is perceived by others, are also qualities that have the potential to easily alienate those on the receiving end."[2]

The takeaway of the Cornell study is that self-awareness is the overall key predictor of success in business. But the implications, I think, go far beyond the realm of business. Self-awareness is worth giving some serious thought to period, especially if you are currently struggling in your life. Becoming aware of your limiting beliefs is the first step in removing obstacles in your life. Increased self-awareness improves communication, team work, personal interaction, and productivity in your personal life, too. The more self-awareness you have, the better your relationships will be with your spouse or significant other, friends, parents, children, etc.

Self-Awareness Should Lead to Self-improvement

One final word on this topic: awareness is one thing, action is something else entirely. Awareness alone will not solve your problems; the reason to focus on awareness is because you can't solve a problem if you don't know what it is. Once you are aware of a weakness or limitation, you have a responsibility to do something about it. If you don't do anything about it, then you are no better off than if you had never become aware of it at all.

1 The study was conducted in 2010 by Green Peak Partners and Cornell's School of Industrial and Labor Relations. https://bit.ly/leaders-self-awareness

2 The study was conducted in 2010 by Green Peak Partners and Cornell's School of Industrial and Labor Relations. https://bit.ly/leaders-self-awareness

This is especially true when it comes to becoming aware of your limiting beliefs. If you are currently unhappy or struggling in your life, it is extremely important that you spend some real time reflecting on who you are, the decisions that you have made, why you have made them, and what this reveals about you and how you have come to view the world. It is also critical that you have this epiphany:

Your view of the world is the way the world is *for you*, because you are creating the world with your vision of it. If you want the world to be different, change your worldview. When you change your worldview, you change your world. When you change your world, you change *the* world.

Your limiting beliefs are simply thoughts. And yet, thoughts are incredibly powerful things. The thoughts that you hold onto, and give power to, tend to manifest in your life. The more you focus on the negative, the more negative things will manifest in your life. You have probably already seen how this works and not realized it. Have you ever noticed that when something goes wrong, as soon as you decide and begin to hold in your mind the thought that you are having "a bad day," more things go wrong? If you are continually reinforcing your limiting beliefs, then your potential will be limited in exactly the way you have imagined it will be. The same thing works in the opposite way. The more you focus on the positive, the more positive things will manifest in your life. As you focus on what you can do and what you will do, a path of possibility will unfold before you.

8

Putting Money Back into the Mix

Nothing in this world can take the place of persistence. Talent will not: nothing is more common than unsuccessful men with talent. Genius will not: unrewarded genius is almost a proverb. Education will not: the world is full of educated derelicts. Persistence and determination alone are omnipotent.
~ Calvin Coolidge

In the last chapter, I gave you some techniques for approaching the thought process that goes into figuring out your why. In the next chapter, I'm going to get hands-on and practical and tell you what to *do*, maybe even *as* you're discovering your why, to start moving in a new direction. But before I do that, I want to bring money back into the conversation, because money is inextricably linked to much deeper areas of our lives than we may commonly believe or understand.

Most people misunderstand money. Some think that money is a goal, an end in itself. Some see it as an obstacle standing between themselves and the life they would prefer to live. Some see the amount of money that they have as a representation of their worth as a person. Some believe that the way to increase money is to hang onto it, when in fact the exact opposite is true.

In previous chapters I told you some important things. I was talking about intangible things like "value" and "happiness," but every principle I shared with you is also true about money. Why?

Because money is also an intangible thing. Money is not a printed piece of paper. As a printed piece of paper, money is worthless, and a dollar bill has no more worth than a hundred dollar bill. As a bit of digital information stored on a bank's server, it is worthless too. We consider money valuable because money as a physical (or digital) entity is a representation of value. Let me say that again:

Money is a representation of value.

As the price of a product, money is a representation of the value of that product. Why does a Lamborghini cost more than a Hyundai? Because it has a higher value or a higher perceived value. To the person that buys the Lamborghini, the Lamborghini is worth more than the Hyundai. If no one thought the Lamborghini had any value at all, then it would be worthless.

As a wage or a salary, money is a representation of the value of your time. A doctor makes more money than a bartender because we consider the doctor's time to be more valuable. Why is the doctor's time more valuable? Because the doctor has invested between 11 and 15 years of his or her life becoming educated about something that always has value to someone, somewhere. As such, it is not the time itself that is valuable, but the combination of the education behind it and the application of that education within the given time. The doctor is not "better" than the bartender. I'm sure we've all had some fascinating conversations with some bartenders over the years and some head-scratchers with a few doctors. The same doctor's time has no more value than the bartender's time if both of them are sitting at home on their respective sofas, watching a football game and eating Cheetos.

But there's a bigger picture to consider with regards to the value of their respective time.

The example of the bartender's time at work being worth less than the doctor's time at work does not depend on what the bartender does or doesn't know. It depends on whether or not and how he is applying what he knows and how he is giving value to the world. Remember what I told you in Chapter 3? Education and college are *not* synonyms. The bartender could have the same medical degree that the doctor has. But if all he is doing with his 8 to 10 hours a day is mixing drinks, then he is not adding any more value to the world than a bartender who is a high-school dropout. As such, the bartender is no different from my friend Mitt at a job fair with his MBA. If Mitt doesn't know what value his MBA has, and he has no clear idea what to do with it, and he doesn't use it, then he might as well not have it. Those three letters behind his name do not alter the value of his time one bit. If he doesn't understand how to apply his education in a way that adds value, you can sure as hell bet no one else he talks to is going to understand its value either. His mistake and biggest risk was simply assuming that having MBA on a resume would land him a job.

But what if the bartender isn't "just a bartender?" What if he's working as a bartender to keep a roof over his head while he works at a startup for free during the day so he can learn about business and marketing from the inside out and get real, hands-on experience? What if the bartender does this for three or four years and then starts his own successful online business?

Could the same bartender potentially make as much money as the doctor with his online business? Hell yes! He could make more! Does he have to invest 11 to 15 years to go to college, then to medical school, and then do a residency in order to make that kind of change in his life? Only if the bartender's passion and purpose in life is being a doctor. Otherwise, hell no! He can achieve the same income stream for himself over a far shorter timeline by getting specific, targeted knowledge and experience relevant to the modern economy.

This is similar to what my buddy Jason Stapleton did after his time in the Marines. He went downrange as a contractor for a few years to save some money for his growing family. But in his free time, he taught himself the art of FOREX trading and grew a following. He then realized he had a skill others wanted to learn, so one day he asked his small group of followers if any of them would be interested in watching him trade live in a private group. Hundreds said yes and happily paid him a few hundred dollars a month for the opportunity. *Boom!* Overnight he created a six-figure income for himself. Within a few years, he turned that into a multi-million-dollar business. Now has a personal branding empire around his podcast Wealth, Power, and Influence that you should listen to religiously, as I do, if you want to really understand what's going on in the world today.

What I am trying to illustrate in this example is that for most of us, the process required to make such a substantial change of life circumstances is much simpler and takes much less time than the process of becoming a doctor. I have been talking to you about the foundation of such a change—understanding your why—and I am going to get more and more hands-on about what to do next the further we get through the book. If the bartender begins to pursue his passion and offer his own unique value to the world, then he will eventually have that 11 to 15 years of experience behind him, and then more, because he will be happy with what he is doing. He will have been earning money far earlier than the doctor began to earn money and he won't have the doctor's massive debt. He will be adding value to the world, and he will want to continue to learn more and more about his chosen field because doing so makes him happy and adds value to his own life. But by no means does he need to have 11 to 15 years of education or experience to get started like the doctor does. Here's another secret for you:

You *already* know a *lot* more about something than someone else does, and there are people out there, right now, who are willing to pay you to teach it to them.

I will come back to this point later.

A Different Way to Look at Money

If you are currently living in apathy, and you want to change your life, you *must* learn to look at money differently than you probably do. In previous chapters, I have shared with you some principles which I have found to be true and actively at work in the world:

When you put value out into the world, value is what you will get back.

As soon as you begin to act according to who you are and what you are about, clarity will open up for you about how you will create the life you were meant to live. When you know who you are, who you connect with, and how you like to serve, everything will fall into place.

If you think that you will fail, you will fail.

If you think that it's not possible, it isn't.

If you think that what you have in your life right now is all that you can ever have, it is.

Every single one of these principles is equally true about money:

When you put money out into the world, money is what you will get back.

As soon as you begin to spend money in ways that support and develop who you are and what you are about, you will draw money towards you.

If you think that you will be broke, you will.

If you think that you can't afford it, you can't.

If you think that the money you have in your life right now is all that you can ever have, it is.

Similarly, you must find and stop reinforcing whatever limiting beliefs that you have about money. You almost certainly have them:

- No one in my family has money.
- You can't make any money as a/an...

- Money is the root of all evil.
- Money doesn't matter that much. It's ---- that really counts.
- You have to have money to make money.
- I'm not good with money.
- There's only so much money to go around.
- You can't have money and happiness.
- The only way to make decent money is to get a "good job."
- Only shallow people care about money.
- I could be happy if I only had enough money.
- If I could only make enough money, then I could do what I really want to do with my life.
- I will never get out of debt.
- It's impossible for people like me to make any real money. The system is set up to work against me.
- If I don't save as much money as I can now, I won't be able to support myself in retirement.
- Spending money is irresponsible.
- It is irresponsible not to save money.
- Fill in the blank with your own limiting belief here.

The truth is, money is both a tool and an inexhaustible (yes, I said inexhaustible) resource. Thinking of money as abundance that is available to you right now is like thinking and acting with love or kindness. It will constantly replenish itself and flow back to where it came from—you.

Veterans have been taught to manage their money, but they are afraid of actually using it, unless it's to buy a new fancy car. As part of your soul-searching on your own journey out of apathy, I recommend that you ask yourself the following questions. Who's in charge, you or your money? Is your money a tool that propels you forward into a life of abundance and purpose, or has your money taken on a life of its own and convinced you that its existence as money is more important than anything you might want to do with it?

Changing course, pursuing your passion, and building the life that you want will probably cost you money. Will it cost you over $500,000 like it did me? Not necessarily, and quite probably not. But it will require you to apply resources in specific areas in order to achieve specific goals. These goals could be concrete, like writing a book or building a website. Or they could be "softer," like learning how to speak in public or how to negotiate. Your goals will be unique to you, and they will depend on what it is you want to do with your life, which will be rooted in your why.

Leverage Your Most Valuable Asset

Every veteran and service-member that I talk to has their own unique configuration of finances, responsibilities, passions, strengths, weaknesses, and limiting beliefs. And yet, they all possess the same incredibly important asset, and they all inevitably overlook or undervalue it. You have this asset too. Surprisingly, few people think of this thing as an asset or use it as one. An asset is something that you own outright which can be used to produce positive economic value. *You* have this asset, too. It's not your military pension (although that is an asset). It's not your house or your car or any other object in your life.

It's your time.

Your time is the most valuable asset that you have, and it's diminishing every moment.

Think about how you spend your time. What did you do yesterday? What was the outcome of specific blocks of your time? Were they a net gain? Or a net loss?

What do I mean by a net loss in terms of time? Here's just a few examples.

Drinking and other forms of intoxication: Not only can these take up significant blocks of time in unproductive ways, they can sometimes rob us of the next hours' or next day's energy and brain power. too. In this case, we haven't just lost productivity during the time we actually spent getting and being intoxicated. We've also devalued a small chunk of our future.

Commuting: What do you do with all the time you are currently spending getting to work and back every day? If you're listening to shock jocks, or jamming to tunes, or letting political pundits get you all stirred up, you're probably wasting your time. Commuting is an awesome chunk of time to use educating yourself with podcasts or audiobooks. Pissing away this time on mindless entertainment is a net loss because you could be using this time to get ahead in some way, and you're not.

Recreational shopping: This activity, too, can take up significant chunks of time, and what do you have to show for it? Some people shop for entertainment. Some shop to distract themselves from a fundamental unhappiness. The thrill of acquiring something new provides a counterfeit of happiness. But that "happiness" does not last, which is why some people seem to move from desire to desire to desire. No sooner have they acquired what they want than they want something else. But most of what they acquire is not an asset, or not a significant one, and it is automatically worth less the moment that they buy it. The latest video game console or a new pair of shoes probably cannot generate income for you—unless you have your own Let's Play or unboxing YouTube channel, which is something else entirely, and I'll get to that later. If you shop for no other reason than that you are bored, then shopping represents a net loss because the outcome of that activity is worth less than what you invested in it—both in time and in money.

Watching television: I'm not talking about watching documentaries because you process information and learn better visually than by reading; or because you have an interest in filmmaking or videography. In those cases, watching television can represent a net gain in knowledge. I'm talking about binge-watching something you've already seen twenty times, sprawled on the coach on a Sunday afternoon.

Your job: Yep, your job. Does the time that you spend at your 9-to-5 represent a net gain for you in life? Here's a clue: do you enjoy getting up and going to work every weekday morning? Or are you just "working for the weekend?" Working for the weekend aside, you may consider your job a "productive" use of your time because it results in your paycheck. But is that a net gain? Consider:

- If you are spending more than you earn (i.e., you have a lot of consumer debt, especially if you have been carrying high debt for years and you can no longer remember exactly what you paid for to accumulate it), then the time you spend at work is not a financial net gain. If you continue to spend more than you earn as you trade your time for money, then you are setting yourself up for having to trade *more* time to earn more money, meaning your 9-to-5 job could be a net loss for you in terms of money *and* time.
- If you are doing the exact same thing at work right now that you did a few months or a few years ago, then the time you spend at work is probably not a net gain in knowledge or skills. If your knowledge and skills are not continually growing, then the value of your time is not growing either. You are setting yourself up for having to trade *more* of your time to earn more money, as opposed to earning more money for investing the same amount of your time, meaning that in this scenario, too, your 9-to-5 job could be a net loss for you in terms of money *and* time.
- Does your job suck the life out of you? Are you so worn out after a day or a week of work that you don't have the physical energy or the brain power to do much in the evenings or on the weekends? If so, then your job is "costing" you much more in time than the 40 or so hours a week you spend doing it. It is devaluing your future, just like intoxication or debt. It's a net loss.

When you make major financial choices like buying a house or a car, do you base that decision on living up to your income? Is it important to you that the outer circumstances of your life reflect your income level? This is not necessarily a bad thing in and of itself. However, if the big expensive house or car is your "reward" for going to your job every day—if you are using them to cover up or compensate for a deeper unhappiness—then there are better things you could be doing with your money. You could be using it as an asset, instead, to move you closer towards a life that would make you much happier, independent of the status quo trappings of success. If you're unhappy and apathetic in your job, and you're using the money you earn there to cover that up and distract yourself from unhappiness and apathy, again, the time you're spending at your job is a net loss.

Not to mention, as I discussed in an earlier chapter, if you have a salaried job, then no matter how much more time you invest, your annual income will not increase significantly. If you get paid by the hour, sure, maybe you can get time and a half for working extra hours, but there are only so many hours in a day, so ultimately, an hourly income is capped too.

Am I saying that you should stop doing the things that you enjoy? Or that you should stop going to work and paying your bills? Not at all. I am saying that you should be doing them more consciously. Again, self-awareness is key. By all means, take care of your responsibilities, and spend your leisure time in ways that you enjoy. As part of an overall happy and productive life, leisure time itself is productive. It is restorative. With this level of awareness, sure, have some beers or binge watch TV. It's how we recharge our batteries. It can be a way to reinforce our connections and spark our imaginations. But leisure activities pursued as a form of distraction or procrastination are a sinkhole. So is a job that is not interesting or fulfilling and that offers you no opportunity for growth. Remember what we talked about in the last chapter: self awareness is key. Look at your life. What are you really doing with it? What areas are currently net gains for you? Which are net losses? What limiting beliefs do you have that are down in the control room, operating all of this machinery?

Taking Control

I repeat, time is your most valuable asset, and it's diminishing. What are you going to do with it? You probably don't have the answer to that right now, and hopefully I've given you a *lot* to think about. Deep thinking is good: do a lot of it. Be brutally honest with yourself. Have the courage to look at yourself and at the circumstances of your life and acknowledge your responsibility for it and the role that you played in bringing about how things are for you today.

All of this thinking should *not* be idle naval-gazing. Nor should it be full of self-recrimination. Remember the power of your thoughts. This is not an exercise in recognizing where you are "to blame." It is an exercise in recognizing where you have previously unthinkingly exercised your power to create your life, and realizing that you still have power to create your life, and that your life can be different if you want it to be, beginning right now.

I ended the last chapter with the following reminder: Awareness is one thing, action is something else entirely. I am going to end this chapter in the same way. Thinking is not the opposite of doing. They can and should be complements to each other. When you take control of your knowledge, you take control of your options. I told you that figuring out your why is a step-by-step process of exploration and course correction. What does this mean in application?

Learn something new, and then do something different. Learn from what worked and what didn't. Rinse and repeat.

You have already taken a great first step by picking up this book.

In the introduction, I told you that a state of autonomy has three distinct pillars, and like anything that is meant to stand on three legs, all three of those pillars must be in place and in a state of equilibrium in order to create a solid and secure foundation. Although I haven't yet named them, I have been talking about the first two pillars from a philosophical perspective this whole time. They are Personal Growth and Identity. Now it's time to get practical. In the next chapter, I am going to talk about what to *do* to go along with what I have given you to think about. My method all comes down to three simple actions that are the very heart and soul of aspiration. Why did we put money back into the mix at this point? Because money is the resource that's going to help you along the way.

Get ready, because this is where the magic starts to happen.

9

Pillar #1: Personal Growth

You can have many great ideas in your head, but what makes the difference is the action. Without action upon an idea, there will be no manifestation, no results, and no reward.
~ Don Miguel Ruiz, The Four Agreements

Who are you? Who are becoming? Are you a such-and-such at so-and-so company? Or are you something else...something more?

Regardless of what you need to keep doing for the time being to take care of your responsibilities, commit right now to doing more than just being. Commit to becoming. Commit to building your skills as you look for exactly how you want to live your passion. Strive to be someone who is living to work, not working to live. And it's important to point out that there's no age limit to doing this.

From this point forward, I ask that you take a step out of your old self and towards your future. It's okay if you don't know exactly how that future will look, what you will do, or how things will work. If you start down the path, those things will become clearer to you. But you must choose to be something more than what you have been. Remember where you have been, and where we have been, up till now, in this book. If you choose to settle, you choose mediocrity. The status quo—a life without the driving force of a why, a life lived in an endless cycle of the same thoughts and decisions without ever asking what it's all for—is a life of apathy.

How to Get Going

Albert Einstein said it best:

We can't solve problems by using the same kind of thinking we used when we created them.

Translation: If you want your life to be different, *you* have to be different.

You're probably thinking, "Thanks, Scott. Easier said than done, right?"

Actually, being a different you is not as complicated as you probably think. Here's why.

Your brain is like a computer that has been storing information your entire life. It records everything that you encounter and experience, and it is constantly going through the old information as new information comes in, making "sense" out of things, and forming conclusions about the world. All of this information as a whole is what makes up your world view, and the output of all of this stored information is your thoughts.

Most people think mostly the same thoughts most of the time. The more you think a particular thought, the more power you give to it. Thinking the same thought over and over and over again is like wearing a path in your mind. And when you have a worn, clear path in front of you, you are extremely unlikely to step off of that path and set off into a dense forest where you can't see where you should put your foot next. This is why limiting beliefs are so powerful. We return to them again and again and again because they are easy and familiar. When we step off the path of a limiting belief and head into the forest instead, we have to deal with thorny bushes and spider webs and roots that trip us up. This is why thinking and acting in a way that is contrary to our limiting beliefs is so difficult and uncomfortable and why, when we give up and step back onto our clear path, we feel better for a little while, even though we know that the path is ultimately leading us somewhere that we do not want to go.

The good news is that changing your life basically comes down to a single, simple thing: Clear a new path by thinking new thoughts.

How do you think new thoughts?

If your thoughts are the result of everything that you have ever been exposed to and everything that you have ever thought before, then the only way to begin thinking new thoughts is to expose yourself to something new. Bottom line: The way that you get going on the path from apathy to autonomy is to learn something.

Step 1: Get Self-Educated

You know what you know. But what else is out there? What are people out there doing successfully? What ideas are already out there, taking shape and opening up opportunities that resonate with little glimmers of things you have wondered, thought, or experienced before? You may be surprised. It's time to familiarize yourself with what is going on in the modern economy.

There is a staggering amount of virtual real estate online. There is also a staggering amount of knowledge and experience. Search for something—anything. Remember when I asked you earlier what kind of work you would do in the world even if you didn't get paid for it? Search for things having to do with that. Is there something about the world that has always pissed you off? Want to be part of the solution? Look for things having to do with that. Have you had transformative experiences in your life that you know are real but that you have never been able to really explain to yourself, let alone to someone else? Look for things having to do with that. Look for things having to do with your deepest-seated beliefs. Look for things having to do with your most stubborn limiting beliefs. It doesn't matter where you start or what you start looking for. What matters is that you *look*. When you find something that interests you, stop and explore. Don't worry about what you might be able to do, or how what you are doing right now relates to earning money. Just explore and learn.

Read websites. Read LinkedIn content. Watch videos on YouTube or listen to podcasts. Follow and engage with someone you respect and admire on Facebook, Instagram, or Twitter. Take webinars. Start with the free ones. There are a million of them. If you find something that really grabs you, then invest a few bucks and pursue it further. Read books and ebooks. When you find a website that interests you that offers a free ebook, sign up, download it, and read it. Some things won't resonate with you. But some things really, really will. Spend some time in consultation with people who have been where you are and who have valuable insight to share. Get a new perspective on yourself. Engage. Observe. Grow.

Pay attention to what the creators and the helpers and the leaders out there are doing. Why are they doing it? How are they doing it? What are they saying? How are they saying it? What are they giving away for free? What are they selling? How are these things related? What is it about them that grabbed your attention? What is it about them that makes their message worth your time to listen to? And feel free to venture off the Internet too. Buy books. Take classes. Volunteer somewhere. Take absolutely any step that is going to move you closer to something that awakens a passion within you.

Aim to take at least one thing away from everything that you invest your time in. But don't just file it away in your mind as a new thought. Implement something new in your life. Try something new. Let what you learned direct where you go next. Take the next step. Then the next. Then the next. Think about how each new thing that you learn adds value to your life. How does it add value? What have you invited into your life that wasn't there before?

The process of self-education is two-sided. You are educating yourself about what's out there. But you're also educating yourself about what's inside of you. Your responses to everything that you learn, and the directions that you allow what you learn to take you, give you clues that bring you closer and closer to understanding who you are and what is possible for you.

Self-directed education brings you closer to your why. Why do I say self-directed education? Because becoming more educated about something that truly interests, engages, and inspires you is not the same as an education that is imposed on you from the outside. We have all been to school. We all know what it feels like to be "taught" something in which we have no interest. That kind of education is worse than boring. It's grueling. It's painful. Being forced to endure it is like being in a prison. The things that make you feel that way are the opposite of your why. But we also all know what it feels like to be inspired by something, to encounter something that makes us say, "Wow! Tell me more!" The beauty of self-directed education is that you are free to pursue the Wow! and leave the boring, grueling, painful stuff behind. In the course of your journey, you may come to a point where you find that you need the boring, grueling, painful stuff – you may see it in a new way that resonates with you. It may become more interesting because you finally see the reason for it, and you may then decide you want to learn more. Or you may note that down as something you will need to understand only at the highest level, so that you can eventually outsource the nuts and bolts of it to someone else.

The big takeaway here: You have to learn something new about the world around you. This is not a simple fact-collecting exercise. You have to enlarge your world. But enlarging your world is like putting an addition on a house. You can't make the house bigger without breaking down part of what was already there. Ask anyone who has ever been through a home improvement project. You can't put on an addition without making a mess. There will probably be setbacks. It will probably cost you more than you budgeted for. There may be a period of chaos that seems to last forever before you turn a corner and you can finally really begin to see the new and improved space taking shape before you and begin to live in it. So what is the breaking in this metaphor? What's the mess?

You must question things that you thought you already knew to be true.

If you don't, then you aren't thinking different thoughts. You aren't becoming any different. At the very least, you must begin to think this new thought with complete conviction:

I can be something other than what I have been.

How much time are you supposed to spend following your interests down information rabbit holes? As long as it takes. When you reach the next stage, you'll know it, because something will grab your attention and not let go.

Step 2: Get Motivated

What does it mean to get motivated?

It's not the same as doing hard work or working hard. Someone who is motivated has a motive: a reason for doing something, although that reason may not always be visible or obvious to others. In other words, someone who is motivated acts according to a why.

The status quo is a kind of force. It's like gravity. It exerts a constant pressure in the form of messaging like this is how things are, everybody is doing this, and this is the "right" thing to do. Any status quo is essentially a limiting belief, and it is extremely powerful because it's not just coming from and being reinforced from inside of you, it's also coming from outside of you and from all around you. Like the Earth's gravity, the force of the status quo keeps everything fastened down and fixed in place. In order to escape this force, you need to generate slightly more pressure in the opposite direction, like a rocket needs enough thrust and speed to escape Earth's gravitational pull. Motivation is the only fuel that can provide that kind of power. Without motivation, you will never achieve your escape velocity.

What's out there, in the space beyond the force of the status quo, is autonomy.

As you encounter new ideas and start learning new things, one of three things typically happens.

What happens to most people is that they grasp a new concept for a minute (day, week, month...) or two until the next distraction shows up, and then they forget about it. Either that, or as soon as the new thing that they have learned beckons them off of their own well-worn thought paths, they abandon it. They choose to remain on the path of least resistance—the status quo and apathy.

Sometimes you will encounter a new idea and you'll say, "This is crazy. That's not what all my teachers and the people on TV told me. I call conspiracy or scam!" Maybe the new idea is true and you just can't (or won't) see it. Or maybe it's total bullshit. This is where you are really going to start to feel the tension between the status quo and your desire to leave it behind. If the new idea tugs on you enough, then dig into it. Find out. You may find out that it is bullshit. Or you may find out that the world that you think you are living in is not the world that *is*.

Which brings us to the third thing that could happen: You'll discover something that will make you say, "That's interesting, and frankly it sounds like a better story than what we've been told. Screw it, let's take the red pill."

This is not the time to worry about your weaknesses. As I already explained, your "weaknesses" are nothing more than areas where you may need to apply a resource other than your own time. Focus on your strengths, and all boats will rise with the tide. Most people get this backwards. Most people quit as soon as they encounter something that they think they cannot do. Motivated people take action. Motivated people make changes. Motivated people take stock of their resources (skills, knowledge, connections, finances) and decide how to position themselves to increase growth in all areas.

Step 3: Get on with It

We have been given far too much "direction" about how we should be living our lives, based on the way people lived their lives decades ago. Don't fall for this. It's not helpful in today's world. A baby boomer's career and life advice derived from the 80s and the 90s *does not work* in the modern economy the way that it did 30 or 40 years ago. It is hurting our generation. The world has changed. Our society is experiencing massive upheaval from all directions and is lacking leadership at many levels. At the same time, we're being distracted by media pundits and outdated government advice. Doing what you are told because it's in someone else's best interest is *not* liberty. But far too few people realize that they aren't living lives of liberty because they're comfortable, and going against the status quo is scary. It's really scary if you don't have any human capital to rely on. Most people in our society rely on a boss and a paycheck because it's all that they have.

Human Capital

Capital is a word used in the business world that means any resources needed in order to obtain what the business needs for the production of goods or services. It usually refers specifically to money or is measured in money, in the form of either equity or credit/debt. You could say that everything that you need to start a business boils down to money, since there is nothing that you can't pay someone else to do. In fact, if you ask most people about what it takes to start a business, they will say money.

The average person, unless they come from a background of self-employed people or business owners, think that it is both complex and expensive to start a business. As an across-the-board truth, especially in the modern world, *thi is a myth*. It's an idea that's leftover from the previous economy that we have been talking about, the post-WWII economy (and what came before that) that created our current status quo. Don't get me wrong.

You may need a lot of money to start a business. It depends on what you're doing. If you're starting a brick-and-mortar business that needs a lot of equipment, inventory, or employees like a gym or a restaurant or a food truck, then yes, you're going to need some capital in the form of money. You're going to have to do the work to either save, get loans, or find investors, but it's 2020, now, people! As I have already talked about in this book and will talk about again in part three, the world has changed.

You can set up and get hosting for a website that enables you to process payments online for just a few hundred dollars a year. You can develop products without ever having to pay for creating or storing or shipping inventory. You can get into a service where you're selling other people's ideas, and you don't even have to invest any money of your own. We're all on Facebook. We understand how social media works. We've seen the advertisements. Those are all businesses leveraging the technology and the social platforms of today. This is the way people operate today to successfully make money, and they're doing it more easily and more cheaply than ever before. If you're afraid of social media, I'll wave as you get left in the dust. Our grandparents felt that way about T.V., too, when they were our age. But I'm sure you've seen a Super Bowl ad. Successful businesses leverage the technologies of the time. We now have the ability to do it on a smaller and more focused scale. Do you think I'm spending thousands and thousands of dollars on T.V. and newspaper advertisements to market this book and my programs? No way! 99% of that audience isn't my target market. But you are: the military community and those who support it. I can find you online, say hi, and offer you my services for a fraction of a fraction of the costs most business pay to market themselves.

So to circle back to where I started here, the word "capital," I like to define the word capital in terms of "resources" because that's all that money is. Money is a tool, a resource, and a representation of value. Money, in and of itself, is not what you need to start a business. What you need to start a business is accumulated value in some form. I'm going to say that again:

What you need to start a business is accumulated value in some form.

Okay, you may be saying, then, "What's the form?"

You. You are the form. You are the resource. You are the accumulated value. You already have the most important resource that you need to start a business, simply because you exist, and you have lived your own unique life.

If you think this is my own woo-woo, rah-rah business philosophy, it's not. The value that you bring to the table in an enterprise is called human capital. In business, human capital is the sum total of all the skills, knowledge, and experience that an individual possesses. It's the accumulated value that is you. If you currently having a job, then you are trading your human capital, your value, for a paycheck. Remember what I said in the last chapter, it's not about your time, it's about the value of the knowledge and experience that you leverage with your time.

A business can have all the financial capital in the world, but it doesn't make a lick of difference without human capital. Businesses succeed or fail because of human capital. Even the ability to drum up more money for operating expenses for a business is an act of an individual expending some form of human capital. For a service business especially, the loss of an individual—which entails the loss of all of their experience, knowledge, network, and customer/client relationships—is the loss of income. In every business, at some level and to some extent, every individual = resource/tool/value = money. This is why you already have the most important resource that you need to start a business: yourself. There is no one else on Earth who can do exactly what you can do, in exactly the same way, that will resonate with exactly the same people as YOU can. Whether you already have substantial savings or you don't have a dime to your name and are tens of thousands of dollars in debt, you possess human capital that you can begin to leverage right now, today, this instant.

The secret is personal branding, and I'm going to get into that in more detail in the last part of the book.

The getting on with it step is about recognizing, developing, and building your human capital. It's about laying the foundation for something that you will be able to rely on eventually, something that is all your own, something that you can do according to your own rules. When you get motivated, it will be because you have found something that resonates with you, something you know that you can do. It will be because you have found a space where you feel like you belong.

No, you won't know everything that you need to know.

Yes, you will probably need to learn new things.

Yes, you will probably need to start small and take baby steps.

Yes, you may even need to work for free as you develop and improve the skills that you need. But who cares? We're talking about you doing the work that you have already said that you would do whether anyone was paying you to do it or not, right? And believe me, if you do it like you mean it, if you give it all the value that you've got, that will get noticed, and you will *not* work for free forever.

Ultimately, the getting on with it step is about *getting on with it,* whether or not you know all you need to know. *Learn* what you need to know. Whether or not you have the quals you need in order to charge money for the value that you provide. *Get* the quals. Work for the price of the glowing reference alone if you have to. Work whether you have the money to set up the online presence that you want to have or not. Start small. Leverage what is freely available to you and build from there. LinkedIn, Facebook, Instagram, Twitter, and YouTube are all 100% free platforms, and one of them is the perfect platform to showcase who you are, who you are becoming, what you do, and how you like to relate to the world. For me, that platform is LinkedIn, and in a later section of the book I am going to go into great detail about what can be done on LinkedIn and how to do it.

Bottom line, there's no point in waiting around to see what happens. You are creating your future right now, even as we speak, with the thoughts and actions that you take today. If you keep thinking the same thoughts and doing the same things, then one, five, and ten years from now you will still be the same person, in the same place, with the same problems, disappointments, and frustrations, doing the same things.

So when you find something that captures your imagination, when you get an idea, as soon as you get motivated, don't wait to take action. No one is going to give you an order to start moving forward. Don't wait for a clear path to unfold in front of you. Step off into the undergrowth and start hacking away at it. One thing will lead to another. A new future will unfold before you, and before you know it, you will be a different person, in a different place, living a different life.

Hold Up...What About Money?

At the close of the last chapter, I told you that I had put money back into the conversation because money is the resource that you were going to use to help you along your path to personal growth. Then I proceeded to talk for almost a whole chapter about how you already have what you need to get started, and you don't need money.

What the hell? Which is it?

It's both, of course.

If you have money as a resource, you are going to want to apply it to your personal growth in whatever way that you can afford. If you don't have any money to spare, then you can educate yourself for free. You may not have a dime to spare, but I'm willing to bet you have a smart phone and you pay that bill every month to keep yourself connected. That's the application of money to the process of personal growth. Your smart phone, and the Internet connection that comes along with it, is your gateway to an unlimited free education. Use it.

If you have some disposable income, then instead of using it to buy a new toy, more clothes, a more expensive car, or a six-pack or bottle of wine every day, use it to buy education. Purchase online courses. Buy books. Get coaching from someone who has already done what you want to do or something similar to it. Subscribe to something that will expand your world.

At some point, as you are getting on with it, you will need to apply money as a resource to something that your business requires. You may need to hire someone to do something for you. Do not believe that anything is cost-prohibitive to you, because it isn't. If you need to do something to drum up a few bucks so that you can do what you need to do, do it. Also know that there are many talented, motivated people out there who are working hard to build their own brands, who are at their own Get On With It phase, who would be happy to do work for you for much less money than you might think, simply for the sake of being able to say afterwards that they have the experience, and (hopefully) for the 5-star review or testimonial from you that will help build their own reputation. You've heard the saying, "You get what you pay for." That's true – sometimes. But the fact is, as I am sure you know from your own personal experience, sometimes you get less than what you pay for, and sometimes you get much, much more.

Recognize that value comes in many, many more forms than simply money. Use money as you need it and as you have it, and if you don't have any money, then exchange what value you do have for money. Begin to participate in the world in this way, and you will realize that there is nothing which is not ultimately possible for you.

Opportunity and Imposter Syndrome

As you begin to get on with it, you will be met with opportunity. Learn to recognize opportunity when it arises. It may not look like you expect it to look, and when you see it, it's extremely likely that your first response to it will be, "I'm not ready."

Get. Over. It.

You. Are. Ready.

You have probably heard of imposter syndrome. Imposter syndrome means "the persistent inability to believe that one's success is deserved or has been legitimately achieved as a result of one's own efforts or skills." *Everyone* nexperiences imposter syndrome from time to time. People who are massively successful, household names, experience imposter syndrome. Take a moment right now to Google "12 Leaders, Entrepreneurs, and Celebrities Who Have Struggled With Imposter Syndrome" and click through the slide show. It will blow your mind.

Imposter syndrome is nothing more than your old thoughts and limiting beliefs rising up to try to waylay you. Don't let them. Think your new thoughts. Keep going. The old thoughts *will* fizzle out and die as you ignore them.

If opportunity has presented itself, then you are ready. Take the opportunity. Fake it till you make it. If you weren't ready enough, then the universe would not have offered the opportunity up to you. Remember what I said before about the discomfort of setting off in a direction other than your cleared and well-worn path? That discomfort, that fear, will kick in here, probably big time.

Deal. With. It.

Whatever has offered itself up for you to do, do it, even if it means working in and through the fear. What you have to keep in mind is that this is a journey. There will be ups and downs and struggles and confusion. Trust yourself as you navigate through it. The fear will subside, and every time it does, it will leave a little bit more confidence behind.

In the military we used to do what we called hot wash-ups, where we would review something that had happened or something that we had done in order to learn from it. Those reviews always focused on what went wrong, and what we should have done, and were geared towards making sure we could do better next time. I take a related approach as I continue through my own journey, which I also recommend to you. Rather than focusing on what went wrong as you try new things, though, ask yourself, what's working? What do you enjoy? What don't you enjoy? What continues to pull you? The important thing is that you are drawn to something and not constantly freaking out about how good you're going to be at it.

It sounds like an oversimplification, but when I think back to my beginnings, years and years ago, I paid someone thousands of dollars to build me a website, even though I had no clear idea of what I was building a website for, and when it was finished, she told me, "Alright, so start blogging." And I thought, Blogging!?!? What am I going to talk about??? I didn't think I had anything worth sharing. I didn't think I even liked writing after remembering slogging through those copious notes my mentor tasked me with. But I started down the path anyway, and now I am producing content all the time. I've even written two books! So you see what can happen in a few short years when you allow yourself to be drawn down a path. You start discovering stuff, and that's how you notice problems that can be solved, and in the process of exploring that, you find the audience that you can serve in solving those problems. It's a process of exploration, and as long as you don't quit, you'll eventually find that thing. This is why they always say that it takes 10 years to become an overnight success.

The takeaway here: *Get. On. With. It.*

To Sum Up

You *can* join the more mobile and innovative approach to life and recognize this moment in human history as an opportunity. Join the creators, influencers, inventors, thought leaders, investors, and entrepreneurs who say to themselves, "I want to put my talents or financial resources to work as a value-producing member of society. I'm bored with just being 'productive' at my job to get a paycheck that I use to entertain myself in my free time. I'm looking for more. I want to become valuable to others. I want to leave a legacy. When I get to the end of my life, I want to know that I have been purposeful with my time."

The three-step plan that I just laid out for you is where the rubber meets the road. If you want to live a life that is different than the life that you are living now, growth is the *only* way: growing the way that you think and growing the way that you act. Otherwise, decades will go by, and you'll always wonder what if or what could have been. But growth is not for the weak or the feint of heart. Change in and of itself is easy. It's our feelings and attitudes around it that can make it hard. As I said, there will be messes and setbacks. To return to my rocket metaphor, there will be turbulence. There will be the discomfort of G forces trying to tug you back into the status quo. But keep going. The quaking and shake-up you feel in what you thought the world was about and what you thought your place was in it isn't failure or loss, and it shouldn't discourage you: it's only you approaching your escape velocity, and what's on the other side of that point waiting for you is amazing.

Part III

Achieving Autonomy

10

Introduction to Autonomy

Here's to the crazy ones. The misfits. The rebels. The troublemakers. The round pegs in the square holes. The ones who see things differently. They're not fond of rules. And they have no respect for the status quo. You can quote them, disagree with them, glorify or vilify them. About the only thing you can't do is ignore them. Because they change things. They push the human race forward. And while some may see them as the crazy ones, we see genius. Because the people who are crazy enough to think they can change the world, are the ones who do.

~ Rob Siltanen

Autonomy means the right or condition of self-government. It means that you decide how you run your life, what you do, and when. Do you have or want autonomy? Autonomy is not the same as the ability to successfully run your life. You can keep a roof over your head and food on the table, and pay all your bills on time, and have the money to buy anything you want when you want it, but that doesn't mean you have autonomy. Think about it this way:

- We cannot say that we "choose" to take a job unless we are also free to choose *not* to take that job or any job at all.
- We cannot say that we "choose" what we do with our time unless we are also free to choose to do something different with our time.

- We cannot say that we "choose" to live in any particular place or in any particular manner unless we are also free to choose to live in a different place or in a different manner.

And so on. What I am getting at is that we cannot have autonomy in our lives unless we also have financial control.

A life of autonomy is the exact opposite of life in the military. In the military, we aren't free to choose what job we are going to do, what we are going to do with our time, or where we are going to live. But we all know that we choose to join the military. At least, we have a choice whether or not we want to serve in the military at this particular point in history. We sign up for the military. We voluntarily hand over control of our lives to the government for a certain period of time. We make this sacrifice for our country and for our fellow Americans. We do it because we believe in the principles at the foundation of our country, and we feel that those principles are worth defending. That's all fine and good. The problem comes when we leave the military.

The military breaks us down and builds us back up into what it needs us to be. Our time in the military changes us: sometimes for the better, sometimes for the worse. We put a lot of who we are on hold during our service in order to be what the military needs us to be. When our service is over, we have to try to remember and reconnect with who we are, and it's hard. Remember what I said in the last chapter about our recurring thoughts being like well worn paths that we continue to follow without thinking much about it? The thought that "I must do what I am told" is at the bottom of every action that we take in the military, and hardly any of us leap off that well-worn path at the point of separation and into the briar patch.

Hell no.

For one thing, the last six months or so before we separate, as we are going through the transition process, people are still telling us what to do. And we are still service-members, which means we are still following orders. Hardly any of us take the time to sit down and think through the fact that some of the things we are being guided to do are *not* in fact orders, and they are *not* about our lives as soldiers. They are about the lives we will have *after* the military. They are about our lives as veterans, and as such, we do not have to follow the military's guidance (and it is only guidance, it is *not* an order) on these things. To be specific: as part of the transition process, the military gives us guidance about what to do with our money post-military and what to do with our time post-military, and most veterans do what they tell us to do.

Maybe we do it without a second thought because the path has been so well worn down for us by decades of veterans who have come before us. Maybe we do it because we are confused and maybe even in a bit of a panic. We have never really given much thought to our finances. We don't know what our options are, and we don't know any better. Maybe we do it because we are acting in accordance with limiting beliefs we have around money or around the "best" or "right" or "only" way to earn a good living. Whatever our reasons—conscious or unconscious—most veterans elect the status quo when they separate from the military. Specifically, they get a salaried position and allow that to define their "new career." They sign up for the SBP and VGLI, roll their TSP into their new 401(k), and they never even realize that they have followed orders in matters where they were not required to and had a choice not to.

So you can see why I have spent so much time in the previous chapters talking about the significance of our thoughts and limiting beliefs. Achieving autonomy is as much about winning a battle over our mental habits as it is about achieving control over our finances. As I have shown, we can be financially set for a lifetime, but if we are afraid to use our money in a way that aligns with what we feel is our purpose in life, then we do not have autonomy. If we earn plenty of money, but we do it by trading our time doing something that makes us miserable, and we are not happy, we do not have autonomy.

What most people don't realize, and what I have spent the first two parts of this book explaining, is that like our military service, autonomy, too, is a choice. It is in our power to choose a life of autonomy. Most of us don't, but all of us *can*.

It is in your power to choose a life of autonomy.

This book is for those who are looking to create autonomy for themselves. It is for those who want to go from a place of limitations and feelings of apathy around those limitations to a place of career progression, uncapped income potential, and choice about how much time you work, how you work, where you work, and why you work. It is for those who want the choice to work for someone else to be a choice, not a compulsion in exchange for a paycheck. This book is for those who see and appreciate and want to participate in the modern times that we live in. If you have read this far, then I am pretty damn sure that this book and autonomy is for you.

In the last section I talked about aspiration. I showed you what the path out of apathy actually looks like, and I told you what it can feel like. I explained that a journey out of apathy through aspiration and into autonomy is a journey of personal growth. It requires you to learn something and to think new thoughts. I also talked a lot about a key concept: value. Aspiration is basically about figuring out what you are going to do to add value to the world. It is about getting motivated to add that value, and taking the first steps towards adding that value. When you start to add value to the world, value returns to you. It returns to you as satisfaction, as happiness, as a sense of time well spent, as a sense of confidence and personal power and purpose, and at some point, yeah, as money. The point at which the value that you put out into the world begins to return to you as money is the point at which you cross over from aspiration into autonomy.

Aspiration is about personal growth. But personal growth does not end. It continues on into autonomy.

Autonomy is still about personal growth because it is about creating and protecting wealth. Not creating and protecting money, but creating and protecting wealth.

Remember, wealth is not about money. Wealth is about controlling the source of your income doing something that you love to do. Wealth is about choosing what you do with your time. Wealth is about having enough money to live the way that you want to live, for as long as you live, regardless of what kind of lifestyle that you choose. It is impossible to put a dollar figure on wealth, because wealth expressed as a dollar figure is different for everyone and depends on pretty much everything *but* money.

Unpacking Autonomy

I've been framing autonomy as freedom, and it is, but what I'm really talking about is self-reliance. Self-reliance means that you build your own life, you clear your own way forward, you solve your own problems, and you create your own future without requiring the intervention of orders or a paycheck from either a boss or the government. Autonomy means that you are 100% responsible for your successes and your failures. It means that when a problem crops up, you have the resources to deal with it without needing to turn to the government or whoever is next up the chain at your job.

I have talked about a lot of things that you need to do when it comes to personal growth in order to reach the point of autonomy, but there's still one more. And really, everything that I have talked about up till this point all comes down to this one: you must take responsibility for everything in your life. You cannot take the reins on your future unless you first take responsibility for your past and your present.

Autonomy begins when you take responsibility—and I mean when you *really* take responsibility—for everything in your past and for your life as it exists right now. That means no more blaming other people. No more blaming events that have happened in your life. No more excuses about why things are the way they are. No more excuses about why they can't be different. Taking responsibility means accepting and embracing that when things go wrong, it's *your fault*. Not in a self-defeatist, throwing-up-your-hands, limiting belief (I suck, I will always suck, nothing good will ever come of this...) kind of way, but in a personal power kind of way: My own thoughts and actions contributed to creating my present situation, and therefore I have the power to change my present situation in any way I want to.

Let me provide some examples from my own life to illustrate what the spectrum of taking responsibility can look like.

Recently, I had a day when I literally got screamed at on two separate occasions by four adult human beings. The first was a dog-walking situation. I was walking along the beach in one direction with my dog off the leash, and this dude came towards me with a monster. He let it pull him all the way to my dog. When my dog snapped, he started screaming at me that my dog should be on a leash. Then two ladies walked by and expressed the same opinion. Yup…my bad. Not much else to say about that. Later, I got a call from an affiliate partner who was upset because a project we had been working on wasn't going to pay out like we had expected. There were a whole lot of "WTFs!!??!" and "Have you ever seen this befores!!?!?" I just sat there calmly and took it, because ultimately, like the beach incident, it was our fault.

The Power of Taking Responsibility

I realized that the business SNAFU wasn't about the company or the person that was getting in the way of our goals to help our client. No—this was us getting in our own way. How? By being greedy. By focusing on the paycheck instead of serving or giving. I'm not talking about charity. I'm talking about producing vs. consuming. Whenever you are that desperate or attached to getting more money, your heart will never be fulfilled. The problem will just get bigger when you have more money, and there will never be enough money to satisfy you.

The one thing I know I've done correctly in my life is to detach myself from the belief that money in and of itself matters. I can now see money for what it truly is: fiat currency. Money is essentially fake. Our money only holds value because we all agree that it does. That's what was so funny about playing Monopoly as a child. In the moment, it seemed very real to get $200 when you passed go or to buy someone's property to make more money. We think we are pretending, but remember the feeling you had when you went bankrupt? It hurt. If fact, it probably made you angry: angry at your friend who made a move just before you; angry at the rules of the game; angry at anything else that would take the blame away from yourself.

What I've discovered in life, much as I did when I was a kid playing monopoly, is that passing the blame *never* works because ultimately, there is never any blame to pass. It is always *you*, the individual, who is to blame. Why? Because as many wise people have told us century after century, we cannot control what happens to us in our lives, but we can control how we react to what happens.

The problems that you have are not the result of what has happened to you. They are the result of what you did in response to what has happened to you.

Everything that happens in life is your own damn fault. Accept it. Life gets easier as soon as you do.

It's just simple logic. Every time I am in a situation that I can't believe is happening, I simply say, "Well, I put myself here. How am I going to avoid this next time? There's nothing else I can do now. What's the next action I can take in this present moment to move forward?"

At around the same time that I had my day where I was getting screamed at from all directions, I happened to catch a very timely podcast from Jason Stapleton. It's funny how the wisdom he expresses on his shows always seems to fit exactly with whatever problem I'm facing at that moment. It's happened way too many times for it to simply be a coincidence. No. It happens because he's been there himself. It happens because, as I explained earlier, when you start down your "right" path, opportunities arise before you which help you to move further along that path.

Jason's podcast was an opportunity for me to really assess what was going on in my life and take responsibility for it. He's taken responsibility for everything that got him to where he is, including the time his business partner stole his multi-million-dollar company from him and sued him in the process. It's a shitty story, and man, did he get screwed. But is he back at home screaming at his friends and broadcasting on social media how he got ripped off? Or about how he deserves money because of the expectation that he's owed something? Nope. He took it on his shoulders and began producing value by sharing what he's best at: helping people create brands and influence.

In this same episode, Jason said something that was truly profound: he's created a success contract with himself that he signs each year. The contract says that his success is 100% up to him. That's it. A personal success contract is an instrument for taking responsibility for your life. It's an acknowledgement that if something isn't going your way, it's your fault. *You* put yourself in that position, just like I did with the dog-walkers. Just like our affiliate partner did with the paperwork problem he had.

Aren't we all forgetting something? We are all *human*. Companies are run by humans, so treat them that way. Don't be a jerk about it just because something isn't going the way you want it to. It's much easier and far more effective to take responsibility and perhaps offer constructive feedback. Screaming about it is a sign of desperation. No one wants to be around that.

So, as you get this far along in your journey out of apathy and into aspiration, as you get motivated and get on with it and reach the cusp of autonomy, the big task that you have before you is to take responsibility for everything that is in your past and your present, and for everything that will be in your future. If you don't do this, then to go back to my metaphor from the previous chapter, your rocket will never get beyond the outer reaches of the status quo.

What Are You Trying to Increase?

What frustrates me about this expectation we all have about how we're supposed to participate in the economy is that it's all about a paycheck. I talk to many self-employed and small business professionals who are constantly in my ear about getting paid quicker. Why? So we can buy something? So we can go on vacation? We got bills to pay, bro'!

How unfulfilling is that? And by the way, whose fault is it that you're in a situation that you need this next deal to happen so badly that waiting another week or so for it is enough to set you off on a tirade? Why are you passing responsibility off to a company or saying the client needs this to be done faster? It's a turn-off. What are you really asking for?

You're asking for increase. We all want increase in our life. The problem is that we fall into the trap of believing that increase is about money, so we create our lifestyles around consumption. In order to have increase we need to buy more stuff, join the golf club, pay for nicer schools, a nicer car, or a bigger house in a better area...

There is a huge problem with how the financial industry uses language. Behind the scenes of any financial institution, they don't call financial advisors "advisors." They call them "producers." They publish a list of everyone in the firm and how much money they "produced" for the firm. This is so misleading. Financial advisors don't "produce" anything at all. They simply move money around and share "knowledge" regurgitated from marketing materials or what their certification says to say. They aren't sharing any information that can't be found on YouTube. Some move more money than others and get paid more than others, but there's no value there. Just learning a set of principles, rules, and strategies and regurgitating them isn't producing. Unfortunately, many industries like the financial industry pay people so damn much money that it causes people to forget that they are easily replaceable. They think they're valuable because they're getting a big paycheck, and it's not their fault they feel this way.

But I'm not in this to sell. If I just wanted to sell insurance or investments, then I could do that anywhere. Screw that. I am not like any other financial advisor with pre-conceived and wildly unrealistic expectations of what a traditional financial planner can offer. No. I am producing something of value. That's a little harder to do and takes more time, and I do it so that those whom I am meant to serve come to me instead of me having to chase people around hoping to get a client.

Money is a false reward, and its not the increase that I am seeking. You may not know it, but you're not seeking financial increase either. You think you are because society tells you so, but financial increase is just part of the post-military retirement myth, and you don't have to agree to that.

The Benefits of Autonomy

Autonomy can and will bring you money, but at the end of the day, it's not the money in and of itself that will change your life. The real benefits of autonomy are:

- The deep sense of meaning that you will have as you begin to live with purpose, according to your why;
- The peace of mind that comes when you finally bring who you are "on the inside" into alignment with who you are "on the outside" and begin to act in accordance with your core values;

- The sense of personal power and security that comes along with a life of self-reliance.

Money just comes along for the ride.

This final section of the book is about autonomy. We are going to begin by continuing our exploration into personal growth. We will start really digging into the nuts and bolts of it, and how it can work, and how you can cross that threshold from a labor-of-love value-add start-up project to actually earning money with your identity. Then we are going to take a good hard look at the options that are available to you for protecting and growing the life that you have built for yourself (financial control).

11

Pillar #2: Identity

Always place your becoming above your current being.
~ Jordan B. Peterson

I have been talking about the first pillar of autonomy: Personal Growth. I talked about personal growth in the context of becoming aware of our own thoughts and conditioning, as well as in the context of learning new things about the world, the modern economy, and ourselves. Personal growth is not a stage or a phase that you pass through on your way to someplace else. If you want a life of autonomy, then personal growth is going to have to be your new, permanent state. In fact, that is the case with all three of the pillars that work together to create a state of autonomy:

- Pillar #1: Personal Growth. The most successful people in any field are constantly trying to become better. Without this quality, you'll never take the responsibility that a life of autonomy demands.
- Pillar #2: Identity. To be successful creating wealth and providing value, you have to first understand who you are and your purpose in the world. You have to understand your why. This understanding is the foundation of the main tool you will use to create wealth: personal branding. We will talk about personal branding in some detail in this section of the book. Creating wealth does *not* mean making

money. It means creating *value*. Wealth is not a single sum of money in the bank. It is a lifestyle and a way of being. Creating value is the ongoing act of doing work that you find purposeful and fulfilling and that improves the lives of others.

- Pillar #3: Financial Control: As I have already explained, assets are about more than money. You are your greatest asset. The financial control pillar is about ensuring you are able to continue to create value in your life and in the lives of others. It is about protecting both your income and the income stream itself. It is also about ensuring that you have the ability (the autonomy) to do anything with that income that you want or need to do, when you want or need to do it.

So, let's talk about Pillar #2: Identity.

First, Provide Value

When you create, you provide value. It's easy to create, my friends. You do it every day, and you just don't capture it. That's the loss. If you can't fathom why things don't go your way sometimes, I'll tell you why: It's because you aren't offering value. You're pushing your agenda onto others, and it sucks. It's not efficient, and clearly it backfires. But what if you are already providing value, or you think that you are, and nothing is happening for you?

Introduction to Personal Branding

You are probably familiar with the concept of the starving artist (or musician, actor, etc.). A starving artist is one who puts all of their focus, energy, and resources into their art work, with the hope or expectation that they will one day "be discovered" and will "make it," at which point the value of their work will be recognized and appreciated, and they will be materially repaid for their struggle and sacrifice.

The idea of the struggling artist has been greatly romanticized, so much so that it can happen that when an artist finally does start to make money, they are sometimes accused of "selling out," that is, abandoning their integrity in exchange for money. But let's get real. An artist, author, actor, or musician hasn't necessarily "abandoned their integrity" in order to make money. It could just be that they have figured out a way to provide value to someone, or to enough someones to move the needle for themselves.

Why do so many people struggle unsuccessfully for so long? And I'm not just talking about people in the arts. I saw the same thing happen to my colleagues when I was a financial advisor. As I detailed in my own story in Chapter 6, it happened to *me*. It happens to people in all kinds of businesses. Is the problem that they suck at what they do?

No. The problem is one or both of these two things:

- Hardly anybody knows who they are; and/or
- The people that do know who they are don't perceive them as any different or any better at what they do than the next guy (they have not identified their own unique value proposition).

The problem is that they have no personal brand.

The American Marketing Association defines a brand as "a name, term, design, symbol, or any other feature that identifies one seller's good or service as distinct from those of other sellers." The key phrase here is as distinct from those of other sellers. A brand is a way to distinguish yourself from others, especially those who provide similar goods or services. A brand communicates who you are, what you stand for, what value your goods or services have to offer to potential customers, and how you are different from your competitors. Branding is a way to communicate directly to your target market and ideal customer. At the same time, a clearly communicated brand lets those whom you are not meant to serve know that they should look elsewhere.

Think about car commercials. There's a reason that a commercial for a minivan shows a happy mom chauffeuring half a dozen rowdy, dirty 12-year-olds home from a soccer game, while a commercial for a sports car tends to show exotic locales or mad driving skills. These two kinds of vehicles have very different attributes that are likely to appeal to very different kinds of people at very different places in their lives. Chevrolet can advertise the Corvette to soccer moms all they want to, they are very unlikely to sell any. They're probably not going to sell any minivans to fifty-year-old, recently divorced men in the throes of a mid-life crisis either.

You have to know whom you are meant to serve.

Mass Markets vs. Niche Markets

The reason that the Internet has revolutionized and globalized commerce is because it has brought the purchasing power of niche markets into anyone's reach.

What does that mean?

Mass marketing means marketing goods, products, or services to people on a mass scale. The premise behind mass marketing is the idea that a particular good or service has the potential to appeal to anyone and everyone. You see mass marketing at work for goods or services that everyone needs or could potentially want: car insurance is a good example. If you own a car, you must have car insurance. That makes the potential market for car insurance absolutely massive. Within that massive market, insurance companies brand themselves. To differentiate themselves from the competition, they spend a lot of money and time trying to get their names linked to certain values within the minds of potential customers. Consider the slogans for the following companies:

- Geico: Fifteen minutes could save you 15% or more on car insurance.
- Progressive: You could save hundreds on car insurance.
- State Farm: Like a good neighbor, State Farm is there.

- Allstate: You're in good hands with Allstate.

There are two main value propositions at work here: Geico and Progressive are trying to appeal to people who are looking for a lower cost, while State Farm and Allstate are trying to appeal to people who are more focused on level of service and coverage. People gravitate towards one company or another based on the values that company communicates and how those values resonate with their own priorities as a customer.

A niche market is a significantly smaller market that has unique needs or preferences or a unique identity that differentiates it from the mass market. For example, I recently read an article about an insurance agent who left his job at a large insurance company that specialized in construction (which many large insurance companies do) to start his own insurance company specializing in writing policies to cover international adoptions. His new company was instantly able to find clients and grow rapidly because he was meeting a very special niche need that no one else in the market was meeting. He was able to successfully link his company's name to the value of "expertise in international adoption." Insurance coverage specific to international adoption is a niche market within the massive insurance market.

No one serving a niche market is going to buy commercial time during the Super Bowl. Then again, they don't have to. Word tends to spread rapidly within a niche market, especially a previously untapped niche market like the field of international adoption. Why? Because when you make your goods or services known within a previously untapped niche market, you are basically saying to people, "Look, here is something that is especially for you." Aligning your goods or services within a particular niche establishes your credibility over others providing more generalized products or services. There are hundreds of thousands, maybe millions of niche markets, and the Internet makes it incredibly easy to reach out and communicate with any of them at any time and to get your products and services to them.

Finding those whom you are meant to serve – figuring out the value of your own unique life experiences, perspective, and skills, and what community that your products, services, identity, and/or values resonate with and why – is called personal branding.

Become Known for Something

The opportunity available to you right now is unlike any in all of human history, and so many of you are leaving it on the table, because you think you have to follow a system. You think there are rules. You think there are expectations out there that you have to have an MBA or a CFP, that you have to have leadership and management skills. But none of that matters anymore. It really doesn't. What matters in our society right now is that you are authentic. The way to be successful is find those whom you're meant to serve and connect with them. But you can't do that unless you are being authentically you, and you can't be authentically you if you're just spouting someone else's shit. Now's the chance to become who you're meant to be. And if you want to become that person, then you've got to build it. You've got to learn how to create influence. You've got to create a unique value proposition that differentiates you from everybody else. It's 100% possible for you to do that, no matter who you are, because you *are* different from everybody else. You already have a personal brand, because you have experiences and a perspective on life that other people don't have.

You don't have to get everyone to like you. In fact, you want to do just the opposite. You want to repel those whom you aren't meant to serve. That's what's happening with some readers of this book. Some people are just so closed-minded and set in their ways that not only will they disagree with me, they may even try to attack me about my ideas. You should welcome the same feedback, because it tells you you're onto something, that you're unique, and that you're not willing to conform. New ideas attract motivated people and repel the lazy and simpleminded. If you're in a business with clients, this is exactly what you want, so you don't end up with clients who aren't serious about what they hired you for.

You have a potential audience who could value your unique experience, and when you use things like LinkedIn, you can find that audience. You can find your ideal customer. There's a lot of people online who love what I say and do. And there's other people who yell at me and tell me I'm the worst kind of veteran because I'm a "self-help guru." Well, I sure as shit ain't no self-help guru. I'm just trying to figure my own damn life out. And I'll tell you what else I'm not: I'm not a "financial advisor." I have experience, and I have some knowledge I'm happy to share with you. But really, I'm just a person with a passion and a belief and a unique value proposition that no other "financial advisor" can offer you. Someone else might be able to sell you the same product or services that I do, but nobody else can be me, so they can't steal anything from me, and I don't worry about competition. I don't care if it's Northwestern Mutual or Edward Jones. They're going to get their clients, but my clients, the people who are only meant for me, they're going to come to me because they won't want to go anywhere else. My Wealth and Liberty Strategy is going to resonate with them. My passion to make them "un-"employable resonates with them.

The people who are drawn to me are also drawn to some version of self-employment. The fact that you have read this far in this book means that *you* are drawn to some version of self-employment, whether you realize it yet or not. Whether you're doing it as a side hustle, or you're already an industry-licensed professional in some service/relationship-based field, working for yourself means that you control the source of your income, and that's power. But if you're not setting up a personal brand, you can't control shit. It doesn't matter if you're a "licensed financial professional" who can work your own hours and make as much money as you want. Statistics show that if you don't have a personal brand, if you don't do something to differentiate yourself from the crowd, you're probably not going to be successful.

Ninety-nine percent of the people on LinkedIn aren't participating. They aren't showing up and adding value to the conversation. And if they do show up, they're usually saying, "Look at me, buy my stuff!" or "Hey, I need a job," or "Hey, do you need a retirement plan?" or "Hey, if you're looking to buy real estate, let me know," or "Hey do you need insurance? I'm your guy."

I see all this and I think, "Dude, what's in it for me?" Nothing. I'm not giving any of these people any of my attention because they aren't offering me anything that I can't get somewhere else.

The key to growing your business is not adding more letters behind your name, it's increasing the amount of value that you offer. It's figuring out how you can give people something that they can't get anywhere else, and then giving it to them and becoming known for that. If you can figure out how to do this, customers will come to you, and you'll have more business than you know what to do with.

Why? Because the more clearly your brand communicates your unique value proposition and who it is for, the less "advertising" that you have to do. We've all seen the famous artists that basically paint a big red square and sell it for a million dollars. There's perceived value in that big red square because it was made by someone famous. I could paint the exact same big red square and no one would give me a dollar for it. The exact same phenomenon happens in every field and in every industry. People choose between similar goods or services because of the perceived value of one over the other. Personal branding essentially comes down to becoming known for something.

If you want to be successful, you must become known for something.

My mentor was successful in the financial industry not because he was great with numbers, or because he sold unique financial products, or because he understood personal finance better than anyone else in the company. No, he was so successful because he was very, very good at socializing, and because he came across as honest and competent in his field. His brand was essentially being a hell of a nice guy that people enjoyed spending time with. You have to find a way to stand out, and it can be hard to do, especially in service industries like finance and real estate. I can tell you from experience that most financial advisors are told to spend their efforts figuring out how to stand out in order to be successful in the field. The whole numbers thing is really something you can learn on the fly. And yet, there's a cruel irony here, because on the one hand a big financial company is telling you that you have to figure out how to stand out, but on the other hand, they have all these rules in place that constrain you from doing exactly that.

You already possess the one thing that you really need to succeed in your business. You do not need a big company's name behind yours to give you credibility or respectability. You do not need letters behind your name. You do not need to follow someone else's big, organized system for how to do things. Not anymore. Not today. In the modern age, it's all about you.

As I explained in a previous chapter, succeeding in the modern economy begins with knowing your why. You have to know who you are. You have to know your own unique purpose and the value that only you can give to the world. Once you have that figured out, or you are at least heading in the right direction, what you need is a modern skillset. I'm going to dig into exactly what that means in the next chapter. But before we go there, I ask you to reconsider a success story that I told you earlier in a slightly different light.

Consider All Your Possibilities

In the first part of this book, I told the story of my friend Pete, the physical therapist. I told that success story when and where

I did to illustrate a point, that if you want to be happy, then the decisions that you make and the path that you follow must align with your purpose. But I didn't tell you quite all of the story. I held something back because I wanted to use Pete's story to make a different point here.

I told you that my friend Pete has always wanted to be a physical therapist, but he doesn't want to be just your run-of-the-mill physical therapist. He wants to provide a unique skillset. For example, I happen to know that he got his 200-hour yoga teacher certification. No PT does that. One day, when he has enough knowledge and experience, he could open his own brick and mortar practice and do it the way he really wants to do it. But opening a practice is tough, especially now. It's also expensive.

If he asked me, and he hasn't, I would tell him that he shouldn't spend a bunch of time and resources trying to build a standard PT practice with brick and mortar. That's hard. But building a knowledge-based business is easy. He could write his own version of How To Be a Supple Leopard like Kelly Starrett did. Kelly went from being a CrossFit gym owner in a single town to becoming a CrossFit celebrity overnight. How? He shared his knowledge! Writing a book is just one of dozens of ways that you can create an income stream from your knowledge. In the next chapter, I am going to get into the Modern Skillset and how you can use these skills to create your own knowledge-based business.

12

The Modern Skillset

My one repeated learning in life: There are no adults; Everyone is making it up as they go along. Figure it out yourself, and do it.
~Naval Ravikant

Do you remember your flip phone? You could text on a flip phone, but it was kind of a pain, and basically they were still just phones, and the only real advantage they had over landlines was mobility. Then the "smartphone" came along. And holy shit did it change everything. Now we are glued to these things. Our world runs on them. They have replaced computers, cameras, flashlights, maps, phone books, the brick and mortar shopping experience, and even landlines for a lot of people. Yet, believe it or not, many of our military service-members and veterans haven't learned how to use a smartphone productively and creatively. Because of the nature of our work in the military, we seldom have our phones with us. They're locked up somewhere because we're in a secure location or in the field or deployed. So we have had significantly less practice with how to be creative on a phone. And what I remember from my time in the military is that when we did have our phones, we were bored and not doing anything in particular. We used them as consumers, to play video games or scroll social media.

Why did I start this chapter talking about phones? Because if you have a smartphone, then you already own all the technology that you need in order to start the journey from apathy, through aspiration, and into autonomy.

The modern skillet is about developing the skills that you need to stop being just a consumer in the world and start being a value producer.

Recognize what Facebook is doing to the consumer. We, as its users, are also actually its product. That's their business model, and most people don't realize that. You probably remember the big blowup a few years back when it came out that Facebook was collecting massive amounts of data on all of their users. Everyone complained about it, but it's still going on, and everybody's still on Facebook. That's how Facebook makes money. It collects data and sells advertising. That data is very, very valuable. That's why they keep making all these bells and whistles to keep your attention—so you don't leave Facebook. LinkedIn, Instagram, Twitter, all of the social media platforms are doing the same thing, and most people don't give it much thought and never move beyond the point of being a consumer in the mix.

You cross to the other side when you realize that all of our eyeballs are glued to these various social media platforms, and that you, too, can grab people's attention for two seconds and say, "Hey, I've got something to share that might be relevant and valuable to you." On the consumer side, you are using social media as an entertainment. But as a producer/creator, you can use it to build a network of like-minded people, because the deep data gathering that these social media platforms are doing means that you have the ability to target and get your message in front of exactly the kind of people most likely to want to hear it.

Anything a business ever did in a live, concrete way can and is now done virtually. Social media is just the modern way of doing what people in business have always done: hand out flyers, run ads, conduct workshops, and sponsor events. Only social media lets you do it better. Why better? Because when you stand on a busy corner and hand out flyers, you're basically throwing crap against the wall. You have no idea whether or not the people you hand each flyer to cares about what it is you're advertising, and most of them probably don't. Sales and advertising has always been a numbers game. If you get in front of a large enough number of people, you will sell products and services. The trick is to get in front of the person that actually wants and needs what you have to offer. The modern economy allows you to do just that, the first time, and skip the throwing-crap-against-the-wall step entirely.

In this chapter, I'm going to talk about the skillset that you need to develop and/or hone in order to succeed in the modern economy. As you will see, most of these skills are not about technology. Technology is only the tool that will allow you to earn a living while pursuing your passion. Technology, like all tools, will be more or less useful to you the better you are at using it. Sure, the modern skillset includes the ability to use technology and the amazing platforms that are out there for connecting with other people. More importantly, though, it includes the things that will make you really stand out from the millions of other people who use the same technology every day: authenticity, the ability to connect, creativity, and whatever skill or expertise it is that you want to share with the world. Before any of that can happen, though, you must have vision. You must have the ability to recognize that the existence of this technology is opportunity staring you right in the face, which brings us to modern skill #1.

Skill #1: Vision

The first skill you need to have to cross over into the modern economy is the ability to recognize that you can be more than just a consumer in it.

I have already made the point that the world has changed radically and rapidly in the past twenty years. We all know what's available to us in the modern world because we use these things every day. We use the Internet for everything, often from our phones. We use social media to connect with friends and family, to keep up with sports, culture, news, and the arts. We use our phones to shop. We're all aware of the changes and we all participate in them. That is, most of us participate in them by playing with the new shiny object, the next new thing, and we forget that just 10 years ago, the concept of an iPhone and what we can do with it was mind-blowing. What you probably aren't doing is using these tools and these new methodologies to make money, to make a living, or to improve the way that you can communicate your message and your value to the world.

The first modern skill you need to develop is to stop thinking of these innovations as conveniences, distractions, and toys, and start thinking of and using them as tools. You don't have to be an entrepreneur or even start a business, but what you do need to do is to be smart and set yourself apart. I teach this to clients every day. This is *real* financial advice because you can use this knowledge to build wealth.

One of the foundational ideas of this book, and of my Wealth and Liberty Strategy, is this:

You are in the perfect job when you are doing what you love and being compensated for doing it at a level that makes it okay for you to do it.

Today, you can make a living doing damn near anything. Twenty years ago, you couldn't. Why? It's not because people or their interests have changed that much. It's because before the Internet, the ability to find those whom you are meant to serve, or the ability to find enough of them, if it was even possible, could be very, very expensive, prohibitively expensive, in fact, for almost any average person. Assuming that a large enough customer base for your business even existed in the United States, it could be difficult to find those people. Unlike the pretty much ubiquitous need for car insurance, how were you supposed to locate and market to people who are into bonsai, soap-carving, or making snow globes? It's one thing if you lived in New York City, but what if you lived in a small town in Alabama or South Dakota? There's no way you could get a very niche, brick-and-mortar business off the ground in a really small town unless that town is a tourist destination. You might be able to launch and sustain a mail-order business by taking out ads in specialty magazines, but advertising is a one-way announcement, it's not a two-way conversation. Advertising is not about connecting, and serving is not about selling. Serving, much like the Internet and social media, is about connecting, and the secret to connecting is modern skill #2.

Skill #2: Authenticity

In the last chapter, I gave you a little introduction to personal branding. Personal branding is all about authenticity. When you are authentic, you are being genuine. What does it mean to be genuine? It means that you truly are what you present yourself to be. You are not one thing on the outside and something else on the inside. This should sound familiar, because I have talked about this concept, what things look like vs. what they actually are, several times throughout this book already:

The status quo life doesn't feel on the inside the same way that it looks on the outside.

The kind of identity crisis that I am talking about manifests itself as an uneasy sense that you are not in the "right place" or that you are not around the "right people" or that you are not the same person on the outside as you are on the inside.

This book is about being honest with yourself and with the rest of the world. It's about bringing who you are on the inside into alignment with who you are on the outside. It's about discovering the best way to express *you*.

One of the benefits of autonomy is the peace of mind that comes when you finally bring who you are "on the inside" into alignment with who you are "on the outside" and begin to act according to your core values.

The journey from the status quo to autonomy can be seen, in a sense, as the process of achieving authenticity.

This concept of authenticity is why you don't need a string of letters behind your name or the backing of a big company to prove your value. You need to have knowledge, and you need to have experience, and if the knowledge and experience that you truly desire involves the letters or the big company then by all means do that. But you don't need to hide or disguise yourself with the letters or the big company, and you don't need to get the letters or the job with the big company and then pretend that you fit in at the expense of your own happiness in order to have a successful life.

You will *never* achieve wealth—that alchemical balance between money and happiness that is best suited to who you are and the life you were meant to live—by pretending to be something that you are not, no matter how much money you manage to accumulate. The modern world is no longer looking for generic bodies to fill business suits and specifically defined company functions. Sure, there are still big companies out there looking for that, but the world at large, and the individuals in that world as consumers, are looking for connection, transparency, shared struggle, and authentic, individualized experiences and the insight that comes from these things, not one-size-fits-all solutions.

I have devoted several chapters of this book to what it means and what it takes to develop modern skill #2. Achieving this skill is both a process and a leap of faith. It's both the work of a lifetime and the work of a moment. Developing and honing this skill is important because without it you will never master modern skill #3, Connection. The reason that the Wealth and Liberty Strategy works, and the reason that today you can make a living doing damn near anything, is because technology makes it possible to find and connect with your tribe.

Skill #3: Connection

Connection is about exchange. It is about communication. It is about honesty and transparency. It is about mutual benefit. Connection demands authenticity. If you aren't being authentic, you cannot connect. Connection is about identity. It is so important, and the people who are good at connecting are so successful, because real, authentic, transparent connection gets right to the heart of who we are as human beings and provides something that no product or service offering in and of itself can give.

Why is that? It's because the ability to connect with others and the desire to do so is hardwired into us. Think about it. What is true human nature? It's not living in a big house in the suburbs, commuting for 45 minutes each way, and sitting in a cubicle all day. It's not advertising or consumerism, accumulation or materialism.

No.

Human society, human nature, is built on tribalism. As a service-member or veteran, this is a concept that you should understand, because those of us in the military are the very few that get to go back to true tribalism, especially when we're on deployment, especially when we're tight in, sleeping night after night in close quarters with our buddies, our platoon, our mates. We have to depend on each other. We know we're valued in that group, and if we don't do our job, somebody else can die.

Modern society as a whole cannot give us the experience of belonging to a tribe. Not unless there's a tragic event like 9/11, where in the blink of an eye everyone realizes there's a crisis, forgets all their differences, and comes together. I'm not saying I want more terrorist attacks, I'm just saying that pulling together during times of crisis is human nature. Human history and human nature invariably and overwhelmingly include conflict. Not just conflict in the sense of war, but the conflict of finding food, dealing with issues, and raising children. A tribe is a small group of people banded together by common social, economic, religious, or blood ties. A tribe shares a culture and a language. A tribe is rigidly structured, but with that structure comes equality.

When we are on active duty, we get to experience that. The problem is, we're not aware we're experiencing it. We're not aware of the beauty of it, of the uniqueness, the fact that, "Oh my God, for even just a small period of my life within this framework of modern society, I get to be a true human again." In the military, you never feel alone in the crowd. We're in the same uniform. We feel a part of something. But when we leave active duty, the challenge of separating from the military is the challenge of having lost that tribe. It's both a loss of identity and a loss of purpose. Finding a new place where we fit in can be really tough, especially if we've been in the military for a long time. We can live in these giant apartments in huge cities. We can see hundreds of people a day and never talk to one of them. We don't communicate. We don't know who we are and who we should communicate with, or how we should communicate. We're strangers to each other, and humans weren't designed to deal with that. We're confused about the value that we provide, and there is no tribe. This is why it is so important to reconnect with who you really are before you separate from the military, because the first thing you have to do when you get out is to find your new tribe.

If you don't know who you are, you're just going to end up following a tribe that you don't belong in and don't provide value to. You may even do this multiple times before finding the right tribe. If you aren't able to actively aid in the survival of the group and know that you're contributing to that, you slowly start to lose your sense of meaning. If you do know who you are, then you know who to look for and connect with: people who share your interests and values, people that can teach you things you want and need to learn and who can likewise learn from you.

Why this sudden dive into anthropology?

Because the concepts of the niche market and the tribe are closely related, especially because remember, the ultimate goal is serving, not selling. The people who are most successful in leveraging the tools that the modern economy has to offer are not successful because they are great salesmen. They are successful because they seek first to connect with other likeminded people.

When I was telling you the story of my own personal journey, I told you that 95% of people who start working in the financial industry wash out. There are some very good reasons why this happens, and none of them have to do with how much these individuals know about finances. The 95% failure rate is the direct result of failure to connect and failure to form relationships. Back when I was working in the traditional financial industry, we had a lot of technological tools, but websites and smartphones and social media were still new. Not only that, but the large companies in the financial industry, who pride themselves on being old and established, have very strict rules about how to advertise. Everything must be approved. You aren't allowed to send out a mass email, or post anything on Facebook or whatever. If you wanted to give a presentation, you had to use an approved script, things like that. The approval process took weeks, and there was always the underlying anxiety that whatever you wanted to do wouldn't get approved.

The result of this was that most people wouldn't bother trying to do anything. There are website companies for financial advisors that you can pay like sixty bucks a month and they slap your photo and your name and phone number on the website, but every individual financial advisor's page has the exact same content and the exact same blog posts as every other financial advisor who buys from that website. Nobody's going to these websites. Wow, what a business. And what a bunch of people who are just wasting their money in order to be no different from anybody else and who are not really being valuable! I guess it's nice to look credible, and yeah, it's a little hack for people who are just not going to bother understanding how to use the amazing tools they have available to them. People spend tons of money on this stuff.

But as I already explained, during the period of my life when I was working in the traditional financial planning industry, I was always trying to educate myself. I started reading books. I started learning about online marketing, and how Amazon works, and so on. The more I learned, the more unbelievable it seemed to me that an industry that makes tons of money and where people work on commission would tie their employees' hands behind their backs and make them all the look the same, that it would prevent them from being genuine, and so prevent them from making any genuine connections. I equate this to what happens when people are coming out of the military. During that period of time, we're all getting trained to look the same and to be the same as we enter the civilian workforce.

You cannot build a genuine relationship with others by being just like everyone else, and building genuine relationships is how you invite opportunities that you're going to enjoy and find meaningful. People who are part of a network hook each other up. When an opportunity comes up and you think, "Oh, man, this would be good for so and so," you pass that on. When you see some product or service that someone in your network is offering that really resonates with you, you buy it. Networking, whether you call it that or not, is how you get a job or a date or a new dog or something to do on the weekend. The point is, connecting with others is important, and it works better with people who you like and want to emulate, people who you want to mentor, or people who have been through what you've been through. You connect by leveraging a personal brand, using social media, persuasion, influence, and digital marketing tools with modern skill #4, creativity.

Skill #4: Creativity

The modern economy requires us to tap into our creativity again. The world continues to change rapidly, and you have to be able to bob and weave in everything that's going on. Whether you're looking to start a business, monetize your hobby, or land your dream job, you're going to need to be able to leverage a little art, a little entrepreneurship, a little engineering...

Imagine you're sitting in that room, getting your retirement or transition briefing. You're still in uniform and getting trained on how to leave the military and become a civilian. It's funny in a way, it's a very military thing to do, to set up a rigid training program on how to be what most people just are without ever giving it a second thought. The transition "system" is a bunch of briefs. Lots of checkboxes. But what if it happened a different way?

Imagine that the briefer comes in and tells you that just five years ago, if you looked at the top 10 desired skill sets from a corporate perspective, it was the typical stuff: engineering, leadership, organization, yada yada yada. But in 2020, all of a sudden, out of nowhere, number three is creativity. What if he asked you to raise your hand if you're creative? Maybe one guy who writes blogs in his free time at night might raise his hand, but most of the people in the room wouldn't. We have a very narrow idea of what creativity is. We tend to think that creativity is only about producing works of art, like drawings, paintings, novels, or films. But creativity is the use of the imagination or original ideas. There is absolutely no field or activity in which a person cannot be creative. The world we live in today is what it is because of human creativity.

When we're very young, we're still very much in touch with our natural creativity. If a kindergarten teacher asked a class full of five-year-olds to raise their hand if they are creative, every hand would go up. We're all creative. It's part of being human. But most of us tend to lose touch with that aspect of ourselves because over time we all learn to fear and avoid many of the things that are inherent in creativity, like courage, experimentation, absurdity, and failure. As we advance in school, we are subject to more and more rules and expectations. We become increasingly aware of social norms and less and less willing to stand out. Effort, ideas, the risks that we take, and our failures do not earn us good grades. If we do try to be creative, then only favorable outcomes are rewarded, and that's the best case scenario. What's most rewarded in most cases as we go through school and as we enter the work force is our ability to blend in, conform, and do what is expected of us in the way it is expected.

Screw all that. We are living in the fourth stage of the Industrial Revolution. This is the Internet Age. The opportunity is immense. We have the ability to communicate in real-time with people all over the world, to quickly develop relationships, to share knowledge, value, and opportunity. We can help others be valuable, and we can have fun! That's something else we knew intuitively as children that most of us have lost, that doing our work and trying out new ideas can be fun! Focus on the impact that you can have. Tap into and continue to develop your true identity, instead of trying to conform to what you think are the world's expectations for you and getting caught up in external BS that nobody can control.

Vision, authenticity, connection, and creativity: these are your most potent weapons in the modern economy. They're the "soft skills" without which your hard skills – your ability to do particular work in the field of your choice and to leverage technology to build wealth – won't make a damn bit of difference. It is only when you marry up vision, authenticity, connection, and creativity to your hard skills that you will be able to move the needle and create financial control and autonomy in your life.

It's funny, in a way. It's a perfect example of another poor DoD decision to take on the responsibility to train service-members on aspects of their personal life. How on earth does the DoD have the knowledge and skill set to train people how to be civilians? It's hilarious. I don't mean to disparage the hard work that the people running the transition programs do. Clearly, knowledge and resources are being shared. But seriously, these same people are either still in uniform or have transitioned directly to a GS position with the government. *They never transitioned into the modern economy,* so what do they know about it? It's not their fault. There's no way they could have the resources to teach what's really going on in the real world.

Skill #5: The Skill or Expertise You Want to Share

Don't forget that whatever it is that you want to base your business (job search, etc.) on is a skill, too. As such, it has value. It is also subject to becoming obsolete if you aren't continually honing and/or evolving it. Every field has its own unique body of knowledge and skill sets, and in order to be successful in your field, you will need to master them. The good news, though, is that you are probably already master of far more that you think right now. Whatever it is that you do with passion, whether you are a hobbyist, an amateur, or a long-term professional at it, you almost certainly know tons more about it than many of the people who will ultimately look to you for guidance or expertise. Start where you are and build, and never stop building.

A Quick Callback

It is these first four skills that you will combine together in an alchemical kind of way to create your personal brand and provide value. As a matter of fact, before moving on to the next part of this chapter, you might want to go back and reread the last chapter about identity and personal branding with these first four skills in mind.

You back? Great!

The rest of this chapter is about your options for making that personal brand and the value you have to offer known to the world.

Skill #6: Leverage Virtual Real Estate

I have included technology last because even though is the backbone of the Wealth and Liberty Strategy, as I have already said several times, technology, like money, is useless in and of itself. Technology is a tool, and the purpose of a tool is to serve the purpose of whoever is using the tool. Using technology as a consumer will not change your life. Accumulating skills for using technology is meaningless if you don't know who you are, you don't have something to say, or you don't know who you should be speaking to, how you should be speaking to them, or why. So while achieving autonomy will require you to master the skill of leveraging a piece of virtual real estate, I repeat, you will not be able to do this effectively until you get a handle on the first four skills.

Virtual real estate gives you the ability to communicate with other people anywhere in the world in real time and at *any* time. Even while you're sleeping! With it, you can quickly develop relationships, and share knowledge, values, and opportunities. You can help others to be valuable. You can find those whom you're meant to serve, those you like and who like you, and stop spending time dealing with people that aren't serving you and who don't share you values. The specific kind of virtual real estate that you decide to use will depend on what it is that you do and what method best allows you to deliver your unique message.

My online space of choice is LinkedIn. It's the platform I feel most comfortable with and the one I know the most about. That's why I am going to go into the best way to use LinkedIn in great detail in the next chapter. Whichever of these tools you ultimately decide to use, you will need to stand out and be more intentional with how you're using it. Yeah, you might even spend a little money to promote yourself. Gosh, how many people are getting business cards with just their name and email on it? That's promotion, but it's not very effective anymore. Virtual real estate is the new business card, and I can teach you how to use it effectively. So before diving into LinkedIn and how I apply my strategies there, let's take a quick tour of the tools you have available to you. I'm going to keep things pretty high level and generic, because the type of technology and/or platform that you choose is not the point. The point is that you use that technology/platform authentically and creatively to connect.

These days, there are all kinds of ways to claim your own little piece of the Internet and set up shop, literally or metaphorically. If you want to connect with other people, regardless of what you do, you have to have an online presence. Think about it. If you hear about something new or interesting, and you want to know more about it, what's the first thing that you do? You Google it (or whatever search engine you use), right? In the modern economy, if you don't have a piece of virtual real estate, you basically don't exist. No matter what your purpose is, and no matter what why is driving it, you have to be online. Period. *How* you establish yourself online is up to you. There are many options, both paid and free, and I'm going to hit the highlights.

Websites

First up, websites. Websites are the paid option. You will need to invest in a domain name and pay for hosting. There are a lot of companies out there offering these services. I am not going to go into all the nuts and bolts of setting up a website here. There is tons of information out there, so if you want to learn more about it, start educating yourself.

If the idea of designing a website intimidates you, know that there are platforms out there that make it very easy to design an attractive and intuitive website yourself without having to do any coding. If you feel like you need a website but you just don't want or don't have time to learn how to build one, then there are plenty of competent freelancers out there who would be happy to do it for you. Websites can be simple, single pages or entire online stores. You can set up a landing page where you can give away a free ebook and collect email addresses to build an email list. You can write a blog or an online newsletter. You're the creative. You can use a website to do damn near anything you want to do.

But by no means do you have to have a website, because there are a lot of platforms that will give you some virtual real estate for free. My personal favorite is Linked in.

LinkedIn

LinkedIn is a networking platform geared towards business professionals of all kinds. People use LinkedIn to connect as employers, employees, service-providers, and business to business (B2B). LinkedIn offers so much functionality that you can use it much as you would a website of your own. You can use your LinkedIn profile like the About section of a website. From LinkedIn you can write a blog, publish articles, and post photographs or videos. Probably the only thing you can't do directly from LinkedIn is sell things. I'm going to talk about LinkedIn in depth in the next chapter.

Facebook/Instagram

I have included Facebook and Instagram together because Facebook owns Instagram, and you can effectively manage them both from a Facebook account if you choose to. You are probably already familiar with Facebook. While LinkedIn focuses on business and professional networking, Facebook is geared more towards personal social networking. On Facebook, you can post announcements or short blog posts, photos, or videos. You can set up a Facebook group where people can connect with each other. You can use Facebook Live to broadcast video in real time. You can also use the Messenger app to connect in real time to people who visit your page. Facebook can be a great piece of virtual real estate if your purpose has to do with creating and/or fostering connections within a community of people.

Instagram is an image-centric social media platform that people access almost exclusively from their phones. An Instagram post features an image and an optional caption. Posts are categorized using hashtags, which help users to find the content they are searching for. If you are doing or selling something that lends itself to being expressed in images, Instagram can be a great tool for reaching people. Although you can connect with people via comments and a direct message capability, Instagram is most effective as a gateway tool for getting people to some other piece of virtual real estate where they can learn more about who you are or what you do, and potentially take some action like subscribing to an email list or making a purchase.

YouTube

YouTube is a platform exclusively for video. You are probably already familiar with it. Content on YouTube runs the gamut from education of all kinds to professionally scripted and produced entertainment to cat videos, from social movements to sheer stupidity. If you are more comfortable talking than writing, or if your purpose is to teach people to do something, YouTube can be a great place to camp out online. Popular channels can be monetized through advertising. There is a comments capability, however, YouTube is not set up for real-time communications.

Podcasts

The age of on-demand radio has created a far bigger opportunity than most people realize. As a consumer of podcasts, you can choose to listen to whatever you want at any moment. With close to a million podcasts on the market already, you can find the topics and people that really resonate with you. But the real opportunity is to become a producer of a podcast so that those whom you are meant to serve have the ability to carry your value around with them whenever they need it. Do you see how powerful that is?

It's never been easier to record and publish an episode. You just need your phone and a publishing app like Captivate. All other barriers to entry have been removed. The only one left is your limiting beliefs that you don't have a unique message to share or that there isn't someone out there who will listen. In reality, you talk to people every day. You give advice to some of these people, who greatly appreciate your perspective on things. In fact, you have probably had a serious impact on someone's life a number of times. Is it not your responsibility to leverage technology to share your unique message with as many as can find you? I've been listening to podcasts since 2005, and only 15 years later have I realized how powerful this medium is.

Amazon

You probably don't think of Amazon as virtual real estate, but if you are going to be selling a book or a product on Amazon, it absolutely is. Your book or product page is a place to communicate what you have to offer with potential customers. If you publish a book through Amazon (I will talk a little more about that in the next chapter) then Amazon also gives you a second piece of real estate, an Amazon Author Page, where you can not only provide information about yourself, but also post photos and videos. Amazon is not set up for two-way communication.

The Total Toolbox

The list of online platforms above represents the major players, but it is not exhaustive. As with everything, you will need to do your own research and make whatever you choose your own. How you use the total toolbox is up to you. You can create a website that sells an e-book that you wrote. You can create content for a YouTube channel by just using your phone and talking into the camera for 10 minutes per episode. You can create a coaching group or just be a consultant. You can create courses that teach people how to do things that you're passionate about or to share things with people who are on a similar path as you that you wish somebody had told you 10 years ago. You can do videos on Facebook or YouTube, do a podcast, or write a blog. But remember what I told you in the last chapter:

If you want to be successful, you must become known for something.

The people who are paid the most in any industry are those who are known for something. Take Dave Ramsey. He's technically not even a financial advisor, but he's known for his financial advice. Take Dr Phil or Dr. Mercola. Now, I'm not saying all of us have to go try to become a celebrity like that. Not at all. In this modern age, you can build an audience of just 100 or 1000 people who have really aligned to your message and really appreciate what you have to offer. You have a unique way that you can help this particular group of people. There's three billion people on the Internet, all over the world. Finding your hundred or thousand is easier than you think. Simply bring them unique value with regards to whatever it is you do.

What if your thousand people each paid you just ten dollars a month to subscribe to your information or service? Do the math. That's a six-figure income. Maybe you could do a weekly live Q&A session about something you know a lot about, like I do about LinkedIn. People come in and ask me questions. I bring up my profile and show them some stuff. Everybody's at a different level. Some people just sit there watching. I have had people go and use my techniques in a slightly new way and then they come back and show me what they did, and I also get to benefit from their cool new idea. So the learning can go both ways. That's part of how participating in this modern economy in this way enriches your life. The fact that you're always learning and getting better at what you do because of your interactions with others is part of the bigger picture of wealth.

It may take you some trial and error, but when you get this right, you will be shocked by the speed at which you can have an impact. First you have to find out what you want to do, how you want to serve, and kind of mold your message. But this process is ultimately so much faster and so much cheaper than getting more certifications or degrees.

The technology is here, friends. You just need to apply it.

But What If I Really Just Want a Job?

If you think that everything that I've been talking about only applies to you if you want to be some kind of entrepreneur, you're wrong. Take another look at some of the key takeaways here:

> *You are in the perfect job when you are doing what you love and being compensated for doing it at a level that makes it okay for you to do it.*

> *If you want to be successful, you must become known for something.*

These concept apply just as well to landing your dream job as they do to becoming the next Dave Ramsey or Dr. Phil.

The traditional job hunting process is very siloed. You have to tweak your resume towards every position that you apply for, and you may have to do this two hundred times. You have to go to job fairs. You dress up so that you look like everybody else, and hand out a resume that looks like everybody else's. It's a grueling process. It's hard, it's discouraging, and it sucks. But finding your dream job doesn't have to be this way.

What if you are already connected on LinkedIn with the people you're going to see beforehand? What if you already communicate with them? What if you have already posted a few videos and they know your face so that when you show up at the job fair, they already know who you are? They're not seeing you for the first time, they're seeing you for the second, third, fourth, fifth, or twentieth time. Then you walk in and they literally go, "Hey! Great to see you! Give me your resume, I'm so looking forward to introducing you to our people."

At that point, whether or not you're wearing a suit doesn't matter.

Abandon the one-size-fits-all approach to job hunting and tap back into your creativity. If your talents are more on the engineering side of things, think of your job search as an engineering problem. Use your math or artistic skills and draw out your own unique approach. If you're entrepreneurial, start solving problems. Whichever way you choose to tackle the situation, if you want to be successful, your journey, too, starts with modern skill #1: Vision. Recognize that technology represents opportunity. You're going to have to learn to use today's technology, because we don't go to war with yesterday's technology, right? So why do so many of us try to enter the modern world as a civilian armed with nothing more than yesterday's technology?

I started this chapter talking about smartphones. Our civilian counterparts have had their phones in their hands 24/7 while we've been in the military, where you just don't have that kind of access to your phone. The ability to use a smartphone creatively is a learnable skill that is not just valuable, it's almost necessary. We have already reached a point in our society where you can hardly communicate without it. Learn to use yours to share the best value that you can bring to an organization or to other people, and you, too, can develop autonomy. Remember, autonomy is less about *what* you do and all about being happy and fulfilled while doing it. You don't have to be or aspire to be an entrepreneur or business owner or even self-employed. Your audience may just be the type of HR people or CEOs you're looking to work for. Regardless of your goals, this book, and my philosophies and strategies, are for you too. You can use these modern tools to get your specific message right in front of the exact person you're trying to get to by going to all these job fairs and learning how to do a resume and how to apply for jobs. We all know networking is the key. A lot of times, the opportunities that present themselves to you depend on who you've met. The task here is to figure out a way to use these tools to think like you are the business of you, because you are. Whether you are looking for the job of your dreams, or you are ultimately looking to sell products or services, you need to be able to market yourself, and when you master the modern skillset, you can.

Tying It All Together

More income is how you get more wealth and more security, and you create more income by creating more opportunities. One of those opportunities for more income can be investing. There are lots of different types of investments. You can do asset-based stuff like getting interest from bank accounts or growth from the stock market. You can get into real estate and get income from renting property. You can start a business. There are a lot of different ways to produce income. That is the goal, one way or the other, okay? You build wealth on top of that when whatever way you select to create income and opportunity resonates with your passion and your purpose: your why.

Now, you can leverage that money to promote yourself, to produce more income, while at the same time saving money for security. It can balance out and produce a lot of value to you because you can use that investment to speak specifically to who you need to talk to. This involves using simple tools, things you use anyway. You're just not using them correctly and to your advantage yet because you don't know how. You have nothing to lose by trying this. I'm going to show you how to take the things you need to have anyway, like a good social media profile (I know they're telling you in your TAPS class that you need this), and turn it into something that ultimately brings you income.

13

Steal My Strategies

Some will, some won't; so what? Someone else is waiting.
~ Jack Canfield

In the last chapter I gave you an overview of the skillset that you need in order to be successful in the modern economy. In this chapter, I am going to give you an overview of how I have applied those modern skills to create my own business. Remember, like all successful enterprises, my business has a purpose that is *not* about money and that *is* built on a why:

I'm here to help veterans live with a purpose and develop happiness in life around that purpose.

That's my why.

The strategies I have used to work towards fulfilling my own purpose are online content marketing, writing a book, and optimizing LinkedIn.

A Quick Callback

I told you in an earlier chapter that I like to get up in front of people and present. I am much more comfortable talking than writing. So how have I managed to produce so much written content, including two books? Easy. I have made the most of the things that I *am* good at, and leveraged the talents of online freelancers in areas that I am *not* so good at. For me, online transcription tools have been a godsend. I do most of my writing with my mouth. I get my ideas out, and flesh them out, by talking. I then turn the recording into a transcript and enlist the services of an editor to shape and clean them up. Bam! I'm a writer.

Remember what I told you in the second part of this book:

Self-awareness is the key to success.

Leverage and maximize the things that you are good at. Learn and improve in areas where it seems well worth your time to learn and improve. Apply resources where learning or doing something yourself is not the most valuable use of your time. Getting out my own unique ideas and my own unique perspective on the opportunity that veterans have before them today was definitely the best use of my time. Fussing with grammar and the placement of commas wasn't, so I applied a resource there. Where you spend your time and where you choose to apply resources instead will be particular to your own strengths, weaknesses, and interests.

That said, let's take a look at my strategies.

Strategy #1: Content Marketing

Content marketing means creating, publishing, and distributing free content online around a specific topic or for a specific audience. Many companies consider content marketing purely as a sales magnet or as part of a sales funnel. They draw potential customers towards their website by supplying content that they're looking for. The content builds the company's credibility with the potential customer, and the theory is that at some point, because that credibility has been established, and because the paid content (or the product or service the company is selling) is even better or more completely fills the potential customer's need or desire, some of those potential customers become actual customers.

So content marketing is a marketing strategy, which means many companies, most, in fact, see it as a sales tool. I don't think of it as a sales tool because I am mission-focused. I think of it as a way to give something valuable to my community because I have something of value that I want to share. Remember the skills that we just went over in the last chapter? The reason that it's so awesome and that it works so well is because content marketing is:

- A way to find and connect with your tribe (vision)
- A way to show up as yourself in the world (authenticity)
- A way to attract and communicate with like-minded people (connection)
- A way to make something your own by trying out your ideas and having fun (creativity)
- A way to share your skill or expertise in whatever area or field you love and feel passionately about (whatever skill or expertise you want to share)

Content marketing is probably the reason that you bought this book. You probably saw me on LinkedIn or Facebook or visited my website. You probably read something that I have written or watched one of my videos. Something must have resonated with you. You obviously felt like I had given you something of value, and because of that, you felt that this book would give you more value, so you bought it. Does that mean that I simply sucked you into a sales funnel in order to get a few bucks out of you?

That is not my intention at all.

If my goal was simply to sell financial products, then that's what I would be doing. I know how to do it. I have done it before, quite "successfully," if you consider dollars the only measure of success. The reason that I had $500,000 laying around that I could invest in learning how to create a purposeful and meaningful life for myself is because I understand how to sell financial products. But that's not why I create and share content with my tribe.

The reason that I create and share so much information for free on my website is because I want to find the people I am meant to serve. I want to find the people who want to do business with me because my unique value proposition resonates with them. I want the people who find me to understand what I am about and what

I have to offer right up front, so that those who aren't interested in it don't waste their time, and they can keep moving closer towards what *does* interest them. I don't want to "sell" anything to people who don't actively want to buy it. I want the people that I do business with to see our transaction as a net gain for them, and I want it to be a net gain for me.

When I was only working for money, every hour I spent miserable, and every dollar I made, was in some ways a net loss because it made my world smaller and bleaker. I want to help you live with a purpose and develop happiness around that purpose. Sure, along the way, you may decide to engage in business with me around some financial vehicles if you believe they are the right solution for you and you truly see value in them. But if you decide not to do that, but you still feel that I have enlarged your life, that I have helped to set you on your true path, that I have contributed to you discovering your purpose, and you are happier as a result of having engaged with me for a time, then my life is enlarged as well, and I am wealthier because of it.

So what exactly is "content" then?

Content can be pretty much anything. It can be informative or entertaining, depending on your purpose and your passion. But it all pretty much falls into one of the following forms.

The Written Word

When the Internet first started to blow up, blogs were huge. Some people got very well known and launched new careers as a direct result of their blogs. Bloggers shared useful information and personal experiences with their communities, usually around a specific subject or topic like parenting, homesteading, shopping, etc. Blogs exploded because they were a way for people to be authentic and creative, to connect with like-minded people, and to share their unique experience, expertise, learning curve, etc. about something that was very meaningful to them or that they felt passionate about. Blogs as a concept are no longer as big or as ubiquitous as they used to be, but the written word online is still huge in the form of articles, how-to's, personal narratives, online magazines and newspapers, etc. Writing is still an amazing way to communicate information, and people read both online and ebooks/print books all the time.

That said, I think writing articles is great, and I highly recommend it as a way to get your message out into the world. But if all you do is write, then you are only reaching a fraction of your potential audience because, let's face it, some people do not read. They can read, and they might read if the subject interests them enough, or if there is no other way for them to obtain the information they are looking for. There are a lot of people out there, though, who just don't enjoy reading, but who may still be ideal connections for you. So how can you reach these people?

Audio and video!

Audio

Audio is the fastest growing medium for consuming content today. The sales of audiobooks continues to rise, as does the popularity of podcasts. This makes total sense since the beauty of audio is that you can listen and learn something, or listen and be entertained, while you're doing something else, like driving, walking, or working out at the gym. If you currently commute to work, think about how much of your time every week you spend in the car. What are you doing with that time? Is it productive time, or is that time a net loss, a sinkhole that frustrates, angers, and drains the life out of you twice a day? Could a portion of this captive audience be your tribe? Don't overlook the value and potential of audio when figuring out the best way to connect and provide value.

Video

Video is one of the most popular ways to communicate today. Video is used in marketing more than any other form of media. It also overshadows the written word as a form of entertainment. On YouTube alone, people today are watching 3.25 billion hours of video every month. Depending on what it is that you do, it may make more sense for you to make videos than it does to write a book. Making a video doesn't have to be complicated. It can be as simple as talking to your phone for ten minutes. If you have something to say and you know exactly who you want to say it to, then simply make a video of yourself saying it, and post it to whatever virtual real estate you have claimed as your own. Do this consistently, while taking the time to connect with others who have similar interests and areas of expertise, and over time you will find an audience. As with everything else, the keys to success are authenticity, connection, and creativity.

Interviews

If you're not sure what to write or record, consider just doing interviews. Interviews are useful for a number of reasons. For one thing, when you share the interview with your followers, you're also promoting whoever it was who interviewed you or vice versa, and in sharing their stuff, you're building a relationship with them. Remember, the number one rule of persuasion is reciprocation. When you post these interviews as a way to share your ideas, you're also sharing their ideas. You're not just sharing your words about something you both care about, or some problem you're both trying to solve, you're also sharing their words, and that's going to resonate not only with your followers, but also with the interviewer. This is how you build community.

Record Your Interviews!

Interviews are content, and doing interviews will help you to generate more content. Not only can you post the actual interview itself, but if you record it, you can turn that recording into a transcript. Now you have words on a page that you can turn into a short blog post, or several short blog posts. If writing isn't your thing, then hire someone to help you. You can also use each individual section where you were talking as the script for a short video explaining some single technique or key point. It doesn't have to be some deep, meaningful thing. No matter how simple it is, it will be eye-opening for somebody. You could potentially do this for each question the interviewer asked you. Interviews are gold when it comes to creating content. If you have an opportunity to do an interview, take it.

Even better, BE the interviewer. This is a powerful way to network. You are much more likely to be able to have a conversation with someone who has a bigger network and more influence than you if you are the interviewer because by interviewing them and promoting them, you're helping them and making it easier for them to produce content, so it's almost like you're doing them a favor.

Images

People love images. Images grab our attention. They can communicate more, and more quickly, than words. If you have a passion that can be shared visually, then by all means do that. All virtual real estate not only benefits from, but demands, images. Nothing is more off-putting online than a wall of solid text. Your smartphone is also a camera, so learn to use it as one. If you're not into taking your own pictures, there are millions of pictures online that you can use for free or license. But images aren't limited to photographs. They can be charts or graphics that illustrate some concept. Use your creativity here if you're inclined that way, or enlist the help of a graphic artist to help define your personal brand. However you choose to do it, and whatever way you choose as your primary means of communication, we are living in a visual world, and you will need to use images at some point to communicate.

What's the Point of It All?

When you think of people "engaging" with your content, you probably think of likes and comments. Likes and comments are great, but they're just general feedback. Engagement begins with you. It begins with you showing interest in your followers, asking them what they think about some topic or creating content that answers some question that someone has asked you. Show them that you are listening to them and getting value from them. And you know what? You actually will. Ask your readers, viewers, or listeners what they think about your topic. If you ask, then you're putting something on them and some people will feel compelled to respond.

This only matters if it happens organically, though, which means that while your goal is engagement, you don't want to be so focused on engagement that you forget your purpose. Your job is simply to be valuable and share your insights. When you learn something relevant to your purpose or topic, share it. Share tips. If you learn something from someone else's content, share that you've learned something and who you learned it from, and link to their content. This is how you build relationships. This is how you build community.

One final thought about engagement: read the comments that people leave for you. Comments are a way for you to discover what's resonating with people and what's confusing them. If someone messages you, respond as soon as possible in a way that gives them value. People may ask questions, and just as with the interview process, the act of answering questions and talking about whatever it is that you do will help to clarify your ideas, not just for the person who had the question, but for you. You will discover new connections between things that you may not have seen before. How? Someone may point them out to you in a comment or a message. These are the kinds of communications that will help point you in the next direction you need to go. This is not only engagement, it's a key aspect of how your journey unfolds.

Strategy #2: Write a Book

No matter what it is that you want to do post military, you don't need a resume, you need a book. It doesn't even matter if anybody reads that book except for you because once you create it and you publish it, you will realize that you have control and that you can do what you want with your life.

Writing a book will get ideas out of your head and out into the world. The process of writing my first book clarified a lot for me. It helped me to understand who I am, where I have been, and what's important to me. It helped to clarify my why. It opened up new opportunities. It pointed me in new directions and helped me figure out where I wanted to go next. It also established me as someone with a unique history and skill set who has something to offer others, and it helped to clarify exactly who those "others" are and why. When you write and publish a book, you are making a statement to the whole world: This is who I am. This is what I know and what I believe. This is what I am about.

Publishing a book used to be hard. Finding a publisher, especially for a new or a niche idea, could be nearly impossible. The process was time-consuming, expensive, and more often than not filled with rejection. Twenty-five years ago it was possible to self-publish, but it cost a lot of money, since you had to have 500 or 1,000 books printed at a time, which means you had to pay those printing costs up front. If you managed to get that far, then your next problem was distribution. How were you supposed to make your book available to readers? You could sell them out of the trunk of your car or at the back of the room at speaking events, but getting prominent display space in local bookstores around the country? *No freaking chance.*

Even worse, in the traditional publishing world, the label "self-published" was a badge of dishonor. It was taken as an indication of poor quality, in both writing and production, before anyone even opened the book. Most people who self-published books twenty or thirty years ago worked with "publishing companies" known as vanity presses that would produce print copies of their book in exchange for a big chunk of change, but that was it. There was no quality control. Many of the books produced in this way were poorly written, unedited, full of errors, and poorly designed. There was also no marketing support and no distribution. There was a reason that vanity presses were called that, and authors who used them could easily get a bad reputation and ruin their chances of ever getting a traditional publisher interested in their work.

Did all of that mean that their ideas were not worth reading? Hell no!

Then along came the Internet, and Amazon, and Print On Demand (POD) technology. Amazon put the Internet and POD technology together and completely changed the world.

With POD technology, a book can be stored in a digital, print-ready file and printed only when someone actually places an order for it. This means that authors no longer have to bear the expense of having 500 or 1,000 copies of their books printed in advance of sales. They no longer have the problem of storing these books. The Internet solved the promotion and distribution problem for authors. Using any of the various platforms available on the Internet, authors can make themselves and the existence of their books known and make them available to be purchased and shipped anywhere in the world.

There are a number of self-publishing platforms out there, but Amazon's Kindle Direct Publishing (KDP) is indisputably the king. With Amazon's KDP, you can publish print or ebooks. It's easy to use. They produce a quality physical product. You can make your book available to anyone with an Internet connection anywhere in the world. They can also make it available for purchase by libraries and bookstores. You get virtual real estate, the Amazon book page, to promote your book. And you can do it all for free!

Amazon has leveled the playing field in the publishing world. Today, the social stigma that used to go along with self-publishing is gone. If you are willing to do what it takes to make sure that the book you have written is good and error free (which means hiring an editor and a proofreader if you have to) and that it has a compelling cover (which means hiring a graphic artist if you have to), then your book can be largely indistinguishable from those produced by traditional publishing houses today. Because oh, by the way, traditional publishing has been moving towards the POD business model too because it makes so much financial sense.

Strategy #3: Learn to Rock and Revolutionize LinkedIn

As I said in the previous chapter, LinkedIn is my social media platform of choice. It is the one that I use and the one I am the most familiar with. You may prefer another platform and that's totally fine, but don't skip this section, because it's not just about LinkedIn. It's about how to use LinkedIn with vision, authenticity, and creativity to form genuine connections with others. I have used LinkedIn in order to find my tribe, and you can leverage the techniques that I use on LinkedIn no matter which platform you choose as your home base.

LinkedIn is a business and employment-oriented online platform that operates via websites and mobile apps. Founded on December 28, 2002 and launched on May 5, 2003, it is mainly used for professional networking, including employers posting jobs and job seekers posting their CVs. As of 2015, most of the company's revenue comes from selling access to information about its members to recruiters and sales professionals. Many transitioning veterans use LinkedIn in their job searches. The things I'm about to tell you are my observations from using LinkedIn myself.

Over the last few years, I have been incorporating LinkedIn more and more into my daily business, mostly in paying attention to the current status of the military transition process. All the gurus, transition coaches, and transition assistance programs say that LinkedIn is something you need to set up. I've been to their seminars, I've watched their videos, and I've read their articles, and the thing is, they keep talking about LinkedIn like it's an online resume. I find that fascinating, because clearly you can just post the resume on the profile, you don't need to write your profile like it's a resume.

Veterans, we need to stop looking exactly like everybody else! So many veterans have the buttoned-up, now-I'm-in-a-suit-with-a-short-haircut look, or, geez, I'm-still-actually-in-the-military-uniform look. So many veterans are trying to convey that they are competent without giving any thought to conveying who they are. Transitioning out of the military isn't about translating skills onto a resume or translating them into civilian speak! This is where we're making our biggest mistake. We're leaving the opportunity for freedom and liberty on the table. We're totally forgetting that we get to make a choice at this moment about what we want to do next. But instead of making that choice, we're still putting our lives into the hands of these recruiters. We're looking at them as our HR connection, the person who is going to find us our next duty station. That's a sad thing. All this leadership potential, all this ingenuity, all this experience is missing the biggest opportunity they will ever have for self-branding.

Frankly, because most veterans are not doing what I'm about to tell you, they're making the whole job search process really hard on themselves for no reason. When you're following the masses, it's hard to stand out. But LinkedIn gives us this magical power to stand out, and it's really easy because nobody else is doing it! And frankly, if everybody else did do it, it would work better. So, in the sections below I am going to share three important strategies for improving your LinkedIn presence so that you can make a better impact on the people that you are trying to reach.

Change Your Mindset

The first thing you need to do is just change your mindset and realize that no matter what you are trying to do—find a job or build a business or community—you have to sell yourself. You are now an independent consultant who has skill sets, and you're trying to get somebody to hire you for your skill set as much as for your income-producing ability. When you start looking at things this way, then you can immediately shift to thinking: I'm already an independent consultant, why don't I turn this into a business opportunity, get all the business write-offs, and get independent contracts to have a lot more flexibility? Let's say you don't go that far, and you still want to get a regular job. You might as well learn how to market yourself and those skills sets anyway. Adding general knowledge about sales and marketing to your skill set puts you light years ahead of your peers, because you'll know how to talk to people when they don't.

Changing your mindset inevitably leads to taking the following actions.

First, don't make your profile about what you've done. Make it about the experiences you've had, what you've learned, who you are, and how you contribute to the world. You're an independent consultant. What's the unique story about you that gave you the unique skill set that you have, that allows you to teach certain people in a different way that resonates for them? To learn more about this, read an article called "1000 True Fans" by Kevin Kelly. Essentially, you're just trying to find those perfect scenarios with the right types of people for whom your unique experiences and messages will resonate, especially if that person is the hiring manager you want to attract. It allows you to manage your professional identity, build and engage your professional network, and access knowledge, insight, and opportunities. That's essentially all personal branding is, and it can work whether you are looking for an employer or a clientele base. But you must be uniquely you, or what's the point?

Second, take the time to recognize that you have something to offer, because everybody does. There's always somebody that can learn from you. Accept that, and then recognize that platforms like LinkedIn give you the power to talk to those people from all over the world. Success is just a matter of time, and it is whatever you define it to be. Make some minor tweaks to a headline, change the way you do your photo, add a cover photo, write a summary that explains who you are, what you do, and why you do it. And if you don't know these things yet, that's fine. Look at other people's examples and add to your own as you learn and meet new people. But make yourself look like a desirable person to hire, not just another person in the crowd.

Connect

Using LinkedIn is basically building out a database of free connections. I used to have to go to coffee meetups and networking groups and little mixers, and everybody's really bad at these things, but they're trying to get drunk so they can talk and maybe make a sale. It's the saddest scene. You can avoid all that by simply looking relatively professional on LinkedIn. Heck, you can even look desperate, and people want to connect with you! All you are trying to do at this stage is to build out a database of other people who are relatively active on LinkedIn. But be smart about it. Don't just connect with anybody, connect with the type of people that you want to speak with or emulate.

Some of them will respond because they are looking for connections too. Try to find people who are also trying to do a database build, especially people like me, who already have a good-sized social presence, because that gives you more access to bigger audiences. Then you just try to figure out which people to engage with.

Aim for 1,000 connections to start. Then 2,000, then 4,000... I'm only at 10,000, but the experience of interacting on LinkedIn now is light years away from what it was when I was at 500. It's easier to connect to others and easier to start a conversation with those I am connected to. Forging connections is the work. You may get into little chats with a couple of people or get on a phone call with a couple people, and eventually an opportunity is going to present itself, probably without even anybody caring about a resume. You can still post your resume, but keep your focus on building out a database and starting conversations.

Engage

Engagement is the trick to building out a database. If you want to work for a certain company, start connecting with all the people in that company. If you know a certain job description, start connecting with people in that job description. Aim to connect with the other people who are trying to build out a social audience because they're looking for ambassadors, right? Become engaged on LinkedIn until you find someone that you want to learn from. Become that person's virtual apprentice. Find out what the person truly needs to grow their influence and become a superfan. They will teach you everything they know and allow you to tap into their network if you can simply develop a relationship. That's the new job interview. It saddens me when everybody gets on LinkedIn and then just lets their profile sit there. They don't engage or interact, and they don't communicate to a level that gets you to the personal communication. Personal engagement is the point.

That said, the way I engage is I write articles. They are an opportunity for others to connect with me. If you interact with something that I have written, that lets me know that you're going to be engaged in learning about the things that I've learned. I'm not your guru. I don't charge for this; that's not my shtick. I'm building a business. And so, yeah, I leverage these connections, these opportunities to teach.

Because I am making an effort to connect, and I am actively engaging with people, I'm probably going to grow my business a lot more efficiently. And that's our whole marketing strategy. We teach it to every business we affiliate with. We teach it to every individual we work with because we want to help them build their personal brand. We don't want to keep them as employees because that doesn't make any sense. I'd rather contract them out because they found a better idea. They're going to remember that they were able to grow because we taught them the real way to tap into income-producing opportunities in the new information age, the sharing economy.

These strategies are not just about the Internet. They work well on other social media platforms and in the traditional marketplace, too. The new economy, and tools like LinkedIn, are decentralizing business. That's giving more people more opportunity to tap into all this wealth. The numbers are all there. You've just got to know that going for one particular job and one salary that's tied to one industry is the riskier option. Because what happens if the next technology comes and that entire industry disappears?

That's it, my three tips for rocking and revolutionizing LinkedIn: Change your mindset. Connect. Engage. If you want to go real deep (and you should), check out my free course at LinkedInLiberty.com.

It all comes down to this:

If you want to be successful, you must become known for something.

When you create, you provide value.

A Quick Callback

Think back for a minute to Chapter 9, Personal Growth. In that chapter, I outlined what I see as the three major steps of personal growth:

- Get educated.
- Get motivated.
- Get on with it.

I put these steps back in front of you again here because after the last two chapters you now have a LOT more context in which to consider these steps. It's great that you have read this far through the book. That falls under the "get educated" bullet. I hope that something that I have said has resonated with you, and that you have felt a spark of motivation from one of the skills, tools, or strategies that I have talked about. That falls under the "get motivated" bullet. The next step, of course, is to get on with it.

Take some action. Pick something and do it. Try.

You might fail, but so what? Learn something from it and try again. Wherever and however you decide to start, you must pick something and take action.

What do I mean by taking action?

- If you have unique experience or knowledge, it's never been easier to share it online. Do that! Decide what kind of virtual real estate is best suited to finding your tribe. Where are they hanging out? How do they like to engage? Go there. Start providing value and connecting. Do it consistently and watch what happens.
- If you are already well-networked, good for you. Your network is a resource and a valuable asset. Leverage it! You can push that harder if you build a unique value into your personal brand. Lean on that resource and build up other valuable skills you can offer.
- If you've been smart financially and not only saved money, but kept it liquid so you can use it, then use it! It's never been cheaper to try business ideas, get expert advice that you can implement immediately, and position your assets so that you can build your autonomy without the fear of constantly being tied to a paycheck. You can always tell yourself that one day you will decide whether or not to move into self-employment full-time. Don't over think it. Just be motivated to grow in at least one area.

If you have a unique skill that can be valuable to someone else, leverage that. Consider running Facebook ads for small businesses. Or maybe you can work for free for someone in order to prove your value.

The point here is that there's no point in knowing your why if you aren't going to do something about it. There's no point in imagining and creating a personal brand if you aren't going to use it, and quite frankly, it's impossible to do it in a vacuum anyway! You have to take some step to get started. You have to take some action and get moving so that you know how and when to course correct. If you wait until you think you have everything figured out, nothing will change, and nothing will ever happen for you.

We are living in the greatest period of opportunity in human history. We are living in a time where wealth and abundance is literally being created out of thin air. Today, in the Internet age, wealth is just everywhere. We have the ability to build personal brands, to network, to communicate, and to find an audience. The choice is yours. What would you rather do, slave away for 40 years before you can decide what you want to do with your time every day, if you're even still healthy enough to do it, or spend three or four years now working extra hard to find something that you can truly master and add value to, so that you can literally transform your life?

14

The Problem with Compound Interest

It's easier to fool people than to convince them that they have been fooled.
~ Mark Twain

Now I am going to talk about some financial concepts that are very important for you to understand in order for you to grasp the significance of the financial vehicles I am going to tell you about. In the next chapter, I am going to talk in depth about taxes and how they can potentially impact you in the future. But first, I want to talk about two very important financial terms: compound interest and assets. These terms are commonly misunderstood. It's important that you understand what both of these terms really mean, because you can't make an informed decision about what to do with your money if you don't understand them. Because there is a lot to unpack here with both terms, I am going to dedicate a chapter to each. We're going to start with the concept that underlies traditional retirement planning, what you may have heard referred to as "the magic of compound interest."

Most Investors Buy into the Lie of Compound Interest

Compound interest is basically interest on interest. What this means is that when money that you have saved and/or invested earns interest each month (or, in the case of money invested in the stock market, grows in value), instead of withdrawing that interest, you leave it invested along with your principal so that the total amount earning interest the following month (i.e., the total principal) is larger than the amount that was earning interest the month before. Thus, your interest (or growth in value) is "compounded."

Financial advisors commonly use the concept of compound interest as an argument for why you should invest as much money as you can in the stock market, beginning as early as you can, so that your money has as much time as possible to grow. Most people don't learn about compound interest, if they do at all, until some sort of financial professional tells them about it, usually in some kind of workshop they're holding for young people that they are trying to sign up as clients. Financial firms have done this for decades because they make money when they invest other people's money. Their earnings come in the form of fees that they charge the investors, typically a percentage (sometimes up to 4%) of the amount of money the firm is managing. It's easy to see, then that the more that people invest, the more money the firm makes.

Over time, as the population has grown, and more and more people are saving in retirement accounts, more and more money has ended up in the markets. Most of this money comes from regular people with modest incomes who go to work every day and throw their little chunk of change into the pot every time they get paid. As of the writing of this book, there is something like 29 trillion dollars invested in the stock market just in retirement plans such as TSPs and 401(k)s. Most of this money is taxable, and most of it belongs to somebody who doesn't understand that. Most people put their money into something that they don't understand simply because they're doing what they've been told to do. An even bigger problem is that most financial professionals don't truly understand how the markets work either. They've just been told that over a long enough period of time, money invested in the stock market grows because of this magical thing called compound interest.

How is it that financial advisors are able to continue to sell this story of compound interest? There's a great Einstein quote about it: "Compound interest is the eighth wonder of the world. He who understands it, earns it; he who doesn't, pays it." He's damn right, of course. When it comes to growing your money, there's nothing better than compound interest. Another great example that always shocks people is this question: Would you rather have a million dollars today, or start with a penny and double your money every day for 30 days? And of course everyone's going, "Oh, man, I'll take the million dollars right now. Hell yeah, there's no way that penny is gonna be worth anywhere near that in 30 days." Well, actually, if you double a penny along with the interest that amount can earn every day for 30 days, by day 30 you will have $10,737,418.24![1] That's the magic of compound interest.

Another story commonly told when trying to convince people to invest in the markets is the one about using the rule of 72 to find out how long it will take an investment to double at a fixed annual rate of interest. The way it works is that you divide 72 by the annual rate of return. So, for example, if you earn an average 7% rate of return every year in the markets, you divide 72 by 7 and that tells you that your money should double in about 10 years. So if you have $100,000 invested today, and you average 7%, you'll have approximately $200,000 in 10 years, $400,000 in 20 years, and $800,000 in 30 years. The power of compound interest is why you always hear that the earlier you start investing in the markets, the better off you will be. Someone who starts investing at age 20, the reasoning goes, can easily be a millionaire by the time they are want to retire some 30 or so years later.

So what's the problem?

1 https://nomadcapitalist.com/2019/05/03/double-a-penny/

The problem is that when it comes to the actual behavior of the markets, it turns out that the simplicity of "the magic of compound interest" is a big fat lie. I just told you that lie in the previous paragraph. Did you spot it? Most people don't. Here it is again:

> Use the rule of 72 to find out how long it will take an investment to double at a fixed annual rate of interest. The way it works is that you divide 72 by the annual rate of return. So, for example, if you earn an average 7% rate of return every year in the markets, you divide 72 by 7 and that tells you that your money should double in about 10 years.

The rule of 72 works for an investment at a *fixed* rate of interest. But the rate of return on money invested in the stock market fluctuates. *A lot*. Not just from month to month, or even day to day; it can fluctuate from hour to hour and minute to minute. A person can be rich beyond their wildest dreams in one moment and destitute the next. I recently saw a story in the news of an active duty Navy sailor who somehow created a million dollars in his Robinhood account, and then lost it overnight on a bad trade.[2] All these ups and downs in value get calculated together and sold to the unwary investor as an "average" annual rate of return of 7% to 8%. But an "average" rate of interest is *not* the same as a "fixed" rate of interest. An average rate of return means that on the day that *you* get ready to retire, you could be the person who's rich, or you could be the person who's destitute.

The Truth About Compound Interest

What actually happens is that markets inevitably experience sudden, steep declines in value, and every time this happens, most investors get upset or confused, like I did in 2008. You may have heard the saying that the real job of a financial advisor is

2 https://bit.ly/Robinhood-devastating-results

to keep their client from making "the one big mistake," which is pulling their money out of the market entirely. A lot of people do this. The market takes a dive, and the average American who's been working and putting their money away into their retirement accounts, and who's relying on the "advice" they've been getting from these experts, gets scared when they lose a bunch of their money. So they do what seems reasonable. They pull their money out of the markets so they don't lose any more.

Remember, financial advisors make their money when they have their clients' money invested, so they sure don't want their clients pulling their money out of the markets. They deal with this problem by telling their clients that if they just keep investing and investing that everything will "work out" because of that "average 7% rate of return" and "the magic of compound interest." But what if you're in your seventies or eighties when the market takes a plunge? Are you going to have those ten or twenty years left for your account balance to recover? Or what if you were planning on using that money for a business idea, or maybe just to help you during your military transition?

The important word here is "average." Markets "average" a rate of return over a long term. An average is a single number taken to be representative of a list of numbers. We all know how you get an average number. You add up all the numbers that you have, and divide it by how many numbers you added, and the answer is the average of all the numbers that you added up. We conveniently "forget" how many really LOW numbers contribute to that "average" 7% return until we actually experience them. This is easy to do, since the markets can show a good rate of return for a decade or so before we have a really significant correction. We had one while I was writing this book, because of COVID-19. On the 28th of February, 2020, the stock markets experienced their biggest crash since the housing bubble burst in 2008. However, unlike in 2008, everyone was about to be locked up in their homes for months and hiding behind masks, so most failed to notice it when the Federal Reserve bank just printed new money to prop the stock market back up again.

So how does "the magic of compound interest" work out to investors' favor when this happens?

It doesn't.

Let's look at a simplified example. Say you have $100 in the stock market, and the value of your stock drops 50%. You're left with $50. In the following year, as the market rebounds, market values go back up 50%. Do you have $100 again? No, you have $75, because 50% of 50 is 25. What this means is that the "average rate of return" doesn't equate to how the actual dollars work out. Most financial advisors kind of gloss over this point. They talk around it by talking about things like "using multiple asset classes to balance your portfolio" and "diversification." Or they move portfolios back and forth, chasing higher returns and trying to stay one step ahead of losses. There are a lot of gurus on Wall Street who play this game, making big money moves with other people's money. It's not their money, so it's not really their risk. Yeah, they might lose their job if things don't work out, but the potential of what they can earn if they succeed is massive.

The majority of people with money in the stock market, though, aren't gurus or day traders or financial "professionals." Most people are invested in index funds and are really just along for the ride. They're mystified when their accounts don't just magically go up every single year. Then, instead of making decisions based on what they want for their lives, they make decisions based on what the stock market is doing. They decide to wait it out. Or maybe they reduce their lifestyle so they can contribute more to their 401(k)s in hopes of earning back what they've lost by the time they want to retire. Some people stay invested even when they retire. What happens if they're not protected and there's a crash? Are they going to go back to work at age 75 or 80?

Making decisions based on the behavior of something that is completely outside of your control is the exact opposite of autonomy. And yet, this is the status quo option when it comes to retirement planning. This is the so-called "right thing to do."

There is definitely money being made in the stock market, but the way it's done is to be a trader. You take profits and you take losses, hopefully more profits than losses, depending on how good you are. But there's not that many people who are very good. People who are just passing their money into retirement accounts are competing against the Wall Street big boys, the experts, the ones with huge amounts of money. This group can literally swing markets however they want. They just invest billions of dollars and swing things in their own favor and execute something on it, which is something you can't do if you've just got money in a 401(k) that you don't look at for 5 or 10 years.

The point I want to leave you with is that you should always put your money into things that you understand. This is so important, I am going to say it again:

You should always put your money into things that you understand.

If you don't really understand the markets and what's going on in them, why would you put your money there? Why not put your money into something that offers protection against losses, so you don't have to freak out when the market fluctuates? I am going to talk about what that something is a few chapters on. But for now, having dealt with Compound Interest 101, let's talk about assets.

15

The Truth About Your Assets

If you have built castles in the air, your work need not be lost; that is where they should be. Now put the foundations under them.
~ Henry David Thoreau

An asset is any resource that you own or control that can be used to produce economic value. This seems pretty simple, and yet most people misunderstand it. The problem is that they get hung up on the "owning" part. For example, most people who own their own home consider it an asset. But owning your own home is rarely an asset, because in most cases the home you own does not produce economic value for you. The home that you buy for yourself to live in is not an asset, it's a product. A product is something that has been manufactured for sale. A product is an asset for the person or company that produces and sells it, not for the person who buys it. This is a very important distinction and something that we will come back to later.

Asset vs. Product: An Example

Why is the home you buy for yourself not an asset for you? First, if you are going to see any gain from owning the property it will only be at the point at which you sell it. There is no automatic income stream that comes along with owning the home that you live in. Even if you rent out your basement or the room above your garage, there is no positive cash flow there. You're paying out more every month in your mortgage than you're bringing in.

Second, most people won't see a net gain when they sell their house, either. They will put far more money into their home over the period of time that they live there than they ever get back out of it. Few realize this. When calculating what they have "earned" from their "investment," they don't look beyond doing math at the surface level. If they buy a home for $250,000 and sell it for $300,000 twenty years later, they think, "Great! I made $50,000!" But that's not very accurate math.

For one thing, real estate transactions are complicated, and the purchase price of a home is only one small part of the overall financial puzzle. When you initially complete the purchase, a lot of players are taking a piece of the pie in the form of various fees. This means that right from the get-go your house costs you more than the purchase price. Then, every month, you pay a mortgage, a large chunk of which is interest to the bank. This interest represents an income stream for the bank, since they will ultimately receive significantly more money from you than the amount they loaned you (the purchase price of the house). The home you "own" is actually an asset for the bank that holds your mortgage, not for you. Think about it, what other things that you "own" can be foreclosed on? They are using language intentionally to trick us and make us feel like we accomplished the American dream when really we become a slave to the system.

But what you spend on the home you own doesn't stop there. You also have to pay property taxes and insurance premiums every month. Not to mention that over the course of 10, 20, or 30 years, shit inevitably happens. Roofs, HVAC systems, appliances, and all kinds of other things all have lifespans, and when that lifespan is up, there's no landlord to call to take care of these things for you. When you own a house, you are continually funneling money into it. That's the norm. In the scenario above, the best case would be that the homeowner broke even on the sale of their home; the reality is that they have probably spent more than they gained.

Real estate can also be a lot of trouble. Have you ever owned a property and had your kitchen flood? Or had the market drop right when you're trying to sell? Or been trying to sell or rent a house in another state or country that you are no longer living in? This is exactly what happened to me with my condo in San Diego, and exactly why I will only rent where I live from now on. I plan to invest in real estate again, but why would I invest in something that limits my mobility? Being a "homeowner" often limits your mobility and your flexibility to use your money. It does not give you freedom and liberty. I'm not saying that you shouldn't own property: I'm saying that home ownership should really only be used later in life when you know where you want to stop. Until then, you've got more important and more productive things to do with your time and money. You could earn more income or add more value that will bring you more income later on. I mean, that's just the truth.

Real estate investing, on the other hand, is another animal completely. When you buy real estate as an investment (if you're doing it properly), you go into the transaction with different criteria for the property than if you intend to make the property your permanent home. You don't invest in real estate by hoping a neighborhood maintains its market value as you are paying off the bank and sending property taxes to the government. A real estate investment includes a certain element of control. You don't make the investment without first designing how you will see a positive cash flow from that investment every month in the form of rent or (in the case of house-flipping) a net gain on your investment.

An investment property is (or should be) an asset. The home that you live in is (almost certainly) not. Neither is your car. You may wonder about other things that you have put your money into, things that you can sell if you had to, "investments" like gold, jewelry, wine, artwork, collectibles, whatever. Sure, they may hold value as long as some else still thinks these are valuable in the future. But are they assets? I won't get into it here, but take a moment to Google the Dutch Tulip bubble. They seriously thought flowers were an asset! Ask yourself the test question: Does owning these items provide an income stream for me? The answer is probably no. When you "invested" in these items, much like you probably did when you decided to purchase the home you live in, you may have done so in the expectation that they would rise in value over time. But will they?

First of all, especially over the longer term, you have to consider whether or not any amount that they may appear to have "risen in value" is in fact value, and how much of it is just normal inflation. Secondly, like your traditional retirement account that's invested in the stock market, your ability to see a profit when you sell any of these items depends on their market value at the time you decide to sell them, which means that such "investments" are really just a gamble. Gold is a classic example.

So, if the things that you think are assets actually aren't, you may be thinking that you don't have any assets at all, but that's not true. You probably have at least several assets, you have just never been taught to think of these things as assets, that is, as income streams. In this chapter, I'm going to point these assets out to you. Then, I am going to show you why some of the other things that you have probably considered assets—specifically the status quo financial products that you may be putting your money into right now—aren't usually assets either.

Assets You Probably Have

Your Military Pension Is an Asset

If you are a veteran in the United States right now, you have a pension. If you are still on active duty, you will have a pension. Your pension is compensation for the sacrifice that you have made for your country. It is a way for the taxpayers to pay you back for your service, your time, and the sacrifices that you made of your individuality and your liberty. Your service, your time, and your sacrifices have a unique value that we can actually put a number on. A military pension is usually worth around a million dollars, and can be worth much, much more depending upon your rank when you leave the military. Again, a simple of shift of perspective is required. That 20 years that you gave to your country was a sacrifice. It was also an investment that created an asset which you can now use to your advantage.

Your VA Disability Is an Asset

Some people make even greater sacrifices during their time in service. They might lose a limb, or their sight, or their hearing, or maybe they come home and struggle with PTSD. The government attempts to compensate veterans for those sacrifices with an additional, pension-like payment known as VA Disability. If you receive disability payments from the VA, this, too, is an asset.

Your Personal Brand Is an Asset

Once your personal brand has created income for you in some capacity, it has become an asset that you will need to protect. You don't have to be crazy famous in order for your personal brand to be an asset worth protecting. You don't even have to have your own business. If you're content with your personal brand and the job it enables you to have, cool. Whatever you're doing, if you're comfortable with where you are on an income level, you're going to want to protect that.

I could go on, but I am only trying to show you how to start thinking about the things you have in your life that you currently probably take for granted as assets.

A Quick Callback

In Chapter 7, I went into the concept of limiting beliefs. As a quick reminder, a limiting belief is something that we tell ourselves about ourselves or about the world and how it works. Limiting beliefs shape and limit how we live our lives. We all have them. In that earlier chapter I talked about limiting beliefs in the context of how we see ourselves and our own potential. I referred to them as "insidious little programs (that) often come from our parents or other influential people in our lives, who probably had no idea that they were even installing them or how damaging they could be." Then, in Chapter 8, we looked at some personal limiting beliefs that people tend to have around money.

We all have limiting beliefs about money.

Like limiting beliefs that program our personalities, money-related limiting beliefs can be installed by our parents or other influential people, but they can also be soaked up from the culture that we come from. In fact, the myths around the status quo that I am talking about in this book can function very much like limiting beliefs. Industry norms align largely with limiting beliefs and force people into taking status quo options when it comes to dealing with their money. However, just as a shift in perspective can cause you change how you look at other aspects of how you live your life, like what you do for a living, shifting your perspective and looking at money differently makes the status quo options seem increasingly risky and the more modern strategies make much more sense.

I bring your attention back to this idea of limiting beliefs at this point because as you read the rest of this chapter, you may find your limiting beliefs popping up and whispering in your ear. Simply maintaining an awareness of if and when this is happening will go a long way towards helping you to give the information I am going to provide to you the consideration it deserves.

Before we get into our solution for protecting your assets in the final chapters, it's worth the time to once again revisit the status quo options. This time it's the status quo options for financial planning, and why they don't represent the best option for those who want to become more than average.

Things That May Be Assets (or They May Not)

This book is about achieving autonomy, the "condition of self-government." This means that you are not dependent on anyone but yourself for deciding how you are going to live your life. Autonomy cannot be separated from financial independence. If you can't afford to do what you want or need to do, then you don't have autonomy. If you depend on a job for a steady income stream, then you are at the mercy of your boss, and their boss, and even the state of the economy. Recessions and depressions affect some businesses more than others. So do random and unexpected circumstances like pandemics that can come out of nowhere and shut down large parts of the economy. We just saw this with the coronavirus in 2020. Many businesses were unable to function for months. Some went out of business. So here's a question for you. If your financial independence is dependent on things that are outside of your control, do you really have financial independence? Of course not. You can't be independent if you are dependent.

Let's look at a related scenario. Say you have a million dollars that you can draw on as an income stream, but there's a catch. The catch is that every day when you get up, you have to toss a coin. If the coin lands on heads, your money gains 20% in value. If your coin lands on tails, your money loses 20% in value. Don't forget what you learned in the last chapter, either. If you lose 20% one day and gain 20% the next, you aren't back to where you started. You have sustained a net loss. It's not hard to see that in this scenario it will only take a few unlucky tosses of the coin before you have to start looking for a job.

My coin toss scenario probably seems ridiculous, and yet, if you have all of your money invested in a traditional retirement plan, this is essentially what you are already doing with your money. Of course the gains and losses aren't 20% every day, far from it. But there are gains and losses every day, and sooner or later a day always comes where there is a drop of 10% or

20% or more. People's retirement plans take a huge hit, and they suddenly have to reconsider what they are doing or were planning to do with their lives. As I pointed out in the last chapter when I was talking about compound interest, traditional financial planning and investing is a farce because *no one* has any idea what the stock market will do. When your "financial advisor" tells you not to worry because the market always rebounds, he's just throwing shit against the wall and hoping it will stick. The way that investments are typically sold is misleading, and frankly, the whole system is designed in such a way that it takes 40 years to create autonomy, and that's only if you start in your twenties! Do you want to be 65 years old before you can start living life on your own terms? I sure as hell don't.

You have achieved financial autonomy when you have produced assets and income streams that do not require your constant attention and that do not require you to gamble.

Status quo financial planning says that your traditional retirement plan is an asset. Is it, though, when it comes to achieving financial autonomy according to the definition above? The Wall Street gurus who are making big bucks in the stock market *are* giving it their constant attention, most people who make real money in the stock market (and there aren't many) do. And as far as gambling goes, if all your money is tied up in a traditional retirement plan, then you are dependent on the stock market doing well in order for you to be able to use that money one day. Chances are, you are still decades away from being able to access it without taking massive losses on it in the form of fees and taxes.

In the sections below, I'm going to talk about some of the concerns that I have with some of the financial vehicles that are typically considered "assets." Keep in mind as you read, though, that the reality is that they may or may not turn out to be assets *for you*.

The Problem with Your Traditional Retirement Plan

Most people think of their traditional retirement plan (401(k), TSP, IRA) as an asset.

If you are in the military, and you are part of the new blended retirement system, it is especially important for you to realize that for you in particular your TSP is something that has replaced a true asset—the full retirement pension. Because the government couldn't afford to pay full pensions anymore, they shifted that responsibility over to the service-member by developing a convoluted system that blends a reduced pension with the TSP to better align with the corporate and federal employee retirement systems. This represents a real loss to service-members because now what was a significant guaranteed income stream has become a game of chance. Most don't know how to manage a TSP, and they risk ending up with far less money in retirement savings than they ought to have and far less than they will need.

In general, people tend to think of traditional retirement plans in abstract terms. They invest money every month over decades, that money grows, and at some point after age 59 ½, they have a pool of money from which they can draw an income stream for the rest of their lives. Their money is "invested" and "growing" and will "be there" when they need it.

Well, maybe.

For some people it works out like that. But the devil here is in the particulars. A traditional retirement plan doesn't work out as intended for everybody. The reasons for this are completely outside of your control, which means that funding a traditional retirement plan is not so much an asset as it is a massive gamble. There are two main reasons for this.

The stock market is generally expected to provide an average 7% to 8% rate of return on investments. However, as I discussed in detail in the last chapter, this is incredibly misleading, since this average takes into account massive drops and massive recoveries within the market. Particularly alarming is the sequence of returns risk. During the last few accumulation years or the first few distribution years on a securities portfolio, the owner is only ONE market correction away from financial ruin.

The other wild card is marginal income tax rates. As of April 2020, the United States is $24.22 trillion in debt. Standard and Poor downgraded U.S. credit in 2011, and it has yet to recover. You do not have to be a fiscal or monetary policy expert to understand what that means. When a consumer takes on too much debt and has a reduced credit score, the interest on that debt increases dramatically. That is happening on a national level. Uncle Sam can only do two things to pay down this interest: borrow more and increase taxes. You can expect both to happen in the next 10 to 15 years. This will dramatically increase the taxation on distributions from IRAs, 401(k)s, and TSPs. The fact that the government limits the amount that can be contributed to Roth IRAs (to which you contribute post-tax money) is also telling. The significance of this is that you already have significantly less money in your retirement account than you think you do, because your withdrawals will be taxed, and tax rates fluctuate. What will the tax rate be when it's time for you to retire? It's impossible to say exactly, but there's a real good chance it could be significantly higher than it is right now.

All things considered, when it comes time to start pulling money back out of your retirement plan, the question becomes will you really be pulling an income stream of earned value from the account, or just pulling back out money that you put in which, when all is said and done, after possibly accounting for huge market losses and taxes, isn't worth any more than what you put into it? If that's the case, then it isn't an asset.

When it comes to traditional retirement plans, there are an alarming number of variables outside of an individual's control. To add insult to injury, if you pass away with funds still in a traditional retirement plan, your heirs will pay even more taxes when the money passes to them. The wealthy know how to avoid this kind of loss. We'll talk about how in the next chapter.

The Problem with the SBP and VGLI

As discussed in detail in Chapter 3, the SBP and VGLI are not assets for you; they are costly products that rarely return even the amount that you invest into them. Neither provide much ROI. There are highly improbable circumstances in which both can represent a net gain and become an asset for a surviving spouse, but even if that happens for *your* spouse, you'll be dead. The SBP and VGLI are not and will never be assets for *you*.

The Problem with Term Life Insurance

Term life insurance in the private marketplace does offer some advantages over SGLI and VGLI. Because you have to qualify, the younger you are and the better health you are in, the lower your premium rates and the higher your death benefit are likely to be. However, there is no equity build-up in term life insurance, and like VGLI, the rates get higher the older you get. Unlike VGLI, which will go on for as long as you pay for it, term coverage ends when the term is up. In your sixties or seventies, in the event you are even able to find another 20-year plan, the costs will be astronomical, and you may still very well outlive the term, leaving no protection or legacy for your loved ones. Like SBP and VGLI, term life insurance will only ever be a net loss for you. If there is ever a payment on a term life insurance plan, it will be an asset for your beneficiary, not for you. You'll be dead.

The Problem with Whole Life Insurance

In the 1990s, some financial firms catering to military and veterans began selling whole life insurance as a way to privatize the protection of a military pension, offering a permanent solution to estate planning, something that a 30-year term policy could not do. These were typically offered in conjunction with retirement plans. At a high level, these plans are very attractive. They both provide a death benefit for the insured's loved ones and allow them to leave a permanent legacy. At some point in retirement, the policy is paid up, and the insured no longer has to make payments. They also have the option of continuing to increase their coverage every few years with something called paid-up additions. As long as the insured pays a little more, they are guaranteed to get some more insurance. Over time, taking this approach during a military career was designed to leave you with the perfect amount of insurance to cover the pension so that the SBP would not be needed. While whole life insurance can offer both equity and permanent coverage, which term policies cannot do, these policies have some major flaws:

- Whole life insurance hasn't changed in many, many years. Using these old whole life policies for estate planning today is like choosing to work on an old IBM PC from the 1990s. Sure it performs the same underlying functions, but it is lacking in *decades* of innovation and is simply not the best tool for the job.
- These policies can be very confusing, not only to those that buy them, but to those who sell them as well. Most agents don't really understand what they are selling, so they are not necessarily recommending them because they are the best tool available for an individual client's situation. Rather, these agents are just doing what they're told to do by their agency. Neither do many agents understand the unique situation of military/veterans. To compound the problem, many of the veterans who purchase whole life insurance don't really understand what they are purchasing. It is not uncommon for someone to find out five or ten years into paying for a whole life insurance plan that they can get the same or similar death benefit for a fraction of the cost with a term policy. But comparing a term policy to any type of permanent policy, especially a whole life insurance policy, is comparing apples to

oranges. Veterans who decide that it makes better out-of-pocket-right-now sense to switch to a term policy end up losing what equity they had built into their whole life policy, even if it was only a few thousand dollars.

- Most of the policies that were sold to veterans in the 1990s and since don't offer any advantages for being in good health beyond smoker and non-smoker rates, which makes them essentially no different than the SBP or VGLI in that regard. Furthermore, almost all the costs of these plans are paid for up front, which means that much like a mortgage, it can be a decade or more before you have built any real equity. And although that equity can be accessed tax-free later in life, there is often a loan provision stipulating that you have to pay the company in order to access your money, just like you would get a home equity loan from a bank to access "your" equity in a home.
- Finally, a traditional whole life insurance policy doesn't offer a great deal in terms of equity. The ROI is severely limited by market interest rates. These policies see growth comparable to a bank Certificate of Deposit (CD). That's not a smart way to leverage compound interest over 30+ years. In fact, the way the interest is credited to the equity is in the form of "declared dividends." This is really just the board of executives of the company sitting around at the end of each year and determining how much of the company's profits they want to give back to policy holders, when in reality, those profits came *from* the policy holders, so it's really just a rebate. Furthermore, historically, as more investors move away from whole life, those profits and rebates have been decreasing.

All things considered, a whole life insurance policy *may*end up being an asset for you, decades after initiating one, but there are far better things that you can do with your money that represent much better assets than a whole life insurance policy.

The Best Is Yet to Come

At this point, you may be feeling a little discouraged, especially if you're one of those veterans who has taken whatever financial advice you may have received in good faith and done "all the right things" with your money. But don't worry, as I pointed out in the beginning of this chapter, you have assets. *You* are an asset. In the last chapters of this book I am going to talk about the financial piece of the Wealth and Liberty solution, and how you can leverage it to grow and protect your money so that it can be an asset for you, not only while you are still alive, but sooner rather than later.

16

The Specter on the Horizon: Taxes

For me, it is far better to grasp the Universe as it really is than to persist in delusion, however satisfying and reassuring.
~ Carl Sagan

You probably think of taxes as something that is unavoidable, a necessary unpleasantness that you don't really think about except once a year. Every spring, you either have to make a painful payment, or you receive a refund. And if you're like most people, you probably think of that refund as a windfall or a bonus. (It's not – but I'm not going to get into that here, because that's not what this book is about). Regardless of whether you're writing a check every spring or cashing one, if you're like most people, you probably spend very little time thinking about taxes at all because you consider them unavoidable. If you can't do anything about them, then why bother worrying about them, right?

When I was in Germany, the tax rates there were around 42%. If you made something like 55,000 euros, maybe it was equivalent to around 70,000 U.S. dollars when I was there, you're already in the highest tax bracket. Our tax rates aren't quite as bad here in America (yet), especially right now because of the Trump tax cuts, which are set to expire in 2025. But the first point that you have to understand is that yes, our tax rates are at an all-time low, but tax rates are not permanently fixed. They are adjusted every few years, based on what is going on in our country and what our government needs to pay for. Tax rates are going to increase.

Taxes are a huge expense for most people, and the reality is that those who are on a fixed income and a salary are paying the bulk of the taxes. Why do people always complain that the rich people don't pay taxes? Well, it's because rich people don't just know how to attract wealth, they know how to hold onto it. Rich people own businesses, which allows them to get tax deductions. They understand the rules of this system, including many rules that most people don't even know exist. They participate in that system, and they play by those rules. Rich people don't pay much in taxes because they have some level of financial control, and most people in the middle class and below don't have any financial control. Their money comes in if and when other people decide to give it to them (in the form of paychecks—have you ever been laid off, or had a paycheck bounce, or worked without pay?) and it goes out when the government decides to take it in the form of mandatory withholdings.

So, without doing things like going to Barbados and setting up a tax haven, or setting up some shady shell companies, how do you protect your assets from taxes? And once you've done that, how do you reduce your tax liability in the future? The shorter answer about protecting your money from taxes is to set up things that will give you tax deductions. There are lots of different ways to do that, including real estate, nonprofits, and businesses. People have written plenty of books about those things, so if you want to learn more about that, you can read them. That's not what this book is about.

This book is about is how to protect your money from future taxes.

Protecting your money from taxes that the government will require you to pay in the future is much more difficult to do. For one thing, there aren't that many ways to protect yourself from future taxes. For another, the thing over which you have no control isn't just if or when you will have to pay future taxes at all but how much of your money you stand to lose to future taxes. The rich people have figured out how to put their money in places where they aren't required to pay taxes on it in the future. This is important, since because of some very big problems going on in our economy today, taxes *are* going to be a *lot* higher in the future.

These are not new problems. These same problems were going on back when I joined the financial industry in 2008. People thought things were bad back then. As I write this book, we're now six months into the coronavirus. Many people are out of work with no real prospect of work, and the government is doing things like the Paycheck Protection Program (PPP) and bailing out businesses just so they can pay employees that aren't truly producing because people can't go into their shops or restaurants. See how crazy this is? There is zero value being transferred. Instead of creating value, everyone just sat around and hoped the politicians would save them. You don't want to live that way, do you? And yet, we're all still paying taxes, part of which goes to pay the salaries of these politicians who are making all these stupid decisions. Oh and by the way, you can be sure that the coronavirus hasn't interrupted their income streams one little bit.

So why are taxes going to be much higher in the future? There are three main reasons, and I am going to discuss each one of them briefly in the next sections.

Our Population Demographics Are Changing

The first problem is that we have an aging population. The baby boomers are turning 65 at a rate of something like 10,000 a day. They're going into retirement, and they're living longer. The reason that social security was designed such that you can start drawing money at age 62 is because at the time, people didn't live that much longer after 62. The whole point of paying into social security was that it would help fund your retirement for those last few years you had left to live. But now, with healthcare advances, people are living so much longer that they risk outliving whatever money they did manage to save, which is usually not nearly enough. Compounding this problem is the fact that this aging population doesn't produce. They are no longer members of the working public. They're just consumers. That's the whole point of retirement, the way that we conventionally understand it, right? Retirement means you stop working (producing). It also means that you no longer have a paycheck from which the government can take taxes.

If we look back at the U.S. Census from 1900, 4.1% of the U.S. population was over 65. A hundred years later, in the year 2000, 12.4% of the population was over 65. The projection is that by 2050, 20.2% of the U.S. population will be over 65.[1] Two thirds of all the men and women who have ever lived past the age of 65 since we started keeping records are alive today. Improvements in health care and an overall higher standard of living among a large percentage of the aging population mean that the elderly, retired, consuming population just keeps getting bigger.

Now let's look at what's going on with the younger population. In 1900, 12.1% of the U.S. population was under five years old. These were the people who were going to grow up, go to work, and produce. By 2000, that number was down to 6.8%. The projection is that by 2050, it will be down to 6.3%. Read the book Factfulness, by Hans Rosling, or watch his great TED Talk, where he explains how these demographics work.

What happens with developing countries (we see this in Europe, we see this in Japan) is that birth rates are declining. In some of these countries, the birth rate is below the replacement rate of 2.2 births per woman. America was right at 2.2%, but even the U.S. birth rate has been declining since around 2007. This means that we, too, have a decreasing younger population, which

1 https://www.prb.org/agingpopulationclocks/

https://bit.ly/older-population-grows-dramatically

means that decade after decade, there are less and less working people paying taxes.

The combination of these two trends is financially catastrophic for the government. A lot of retired people have some income in the form of retirement accounts, but not all of them, not by a long shot. The most recent Government Accountability Office report stated, "the percent of households headed by someone aged 55 and over that had no retirement savings decreased from about 52 percent in 2013 to about 48 percent in 2016."[2] Things are heading in the right direction, but over half the people in this country still have no retirement savings. Declining U.S. birth rates over the past few decades also mean that in the coming decades, a lot of those people who will retire at that point won't have kids they can count on for some financial support. So who's taking care of this older population now and in the future? Social security, Medicare, and Medicaid. Somebody's going to have to pay for the aging population, and it's going to be the government, which means it's going to be the working population. If more and more people are needing government support to live, and there are less and less people working and paying the taxes that fund these programs, then there is obviously going to be a shortfall of money, and in fact that is already happening. This is part of the reason why our government is in so much debt.

The National Debt Is Still on the Rise

Before coronavirus, our government was $22 trillion dollars in debt, which was the equivalent of our GDP. I encourage you to study the stats on https://www.usdebtclock.org/. Now, if you and I are in more debt than the amount of income we're making in a year, it makes it kind of hard to qualify for a mortgage, among other things. Yet, our government just keeps taking on more and more debt. The coronavirus stimulus increased the national debt by trillions of dollars. Everyone got $600 or $1,200, and guess

2 https://bit.ly/GAO-retirement-security-report

who ended up with most of that money? The wealthy people who know how to attract it! For example, guess what Walmart did, as soon as they knew the stimulus money was coming out? They stocked up on TVs. There's photos of it on the Internet, all the TVs at every Walmart, because they knew people were going to buy them. So now we're in even more national debt, and everyone's got a new TV and the same financial problems they had before the stimulus.

Unfunded liabilities, like social security, are contributing to the national debt. That's what the term unfunded liabilities means. The government is liable (responsible by law) for paying for these programs, but they can't afford to do it. The money simply isn't there. So the government does the same thing that you or I would do if we had something we urgently needed to pay for and didn't have the money: they get a loan. In some cases, the Federal Reserve simply "creates" more money by adding numbers to a digital balance sheet and then lends it to the U.S. government, who then begins to make interest payments on money that previously didn't even exist, and doesn't really even exist now. Boom. More national debt.

But what does this mean for ordinary people like you and me?

Well, everybody knows that social security is a Ponzi scheme. The money you pay in today isn't being set aside for your old age, it's going to pay the people who are drawing social security payments today. The SBP works the same way. The money you pay into the SBP today is not being set aside for your surviving spouse; it's making payments to someone else's surviving spouse today.

The problem is that this is going to implode for the reasons I explained above. There is no money sitting around to pay for Medicare, Medicaid, or social security. These are unfunded liabilities of hundreds of trillions of dollars. What's the government going to do, just decide one day to stop paying for these things, to stop making social security payments? Are they going to tell everybody, hey, sorry you've been paying into social security your whole working life, but now that it's your turn to start drawing out, we're shutting the whole thing down, and no, there won't be any refunds. Of course they're not going to do that. What they're going to do is pull the money they need from wherever they can get it. They'll just keep taking on more debt, and in order to service that debt, they're going to raise taxes. This won't just affect the working people (of which there are fewer and fewer all the time); it will also affect retirees who are pulling money out of their retirement plans. Unlike the working people, who can work more, or work harder, or work for longer to make sure they can afford to take care of themselves, most of these retirees don't have any other source of income. They can't make up for what they stand to lose as this situation unfolds.

If you don't believe that this can happen, you need only to take a quick look back at our history, and the not-too-distant history at that. U.S. tax rates today are at an all-time low. In the 40s, 50s, and 60s, the highest marginal tax rates were at just about 90%. Sure, very few people actually paid 90% of their incomes to the government, but the possibility of having to give up 40% or 50% of YOUR income to the government sometime in the next few decades is not hard to imagine. It's happened before. It will happen again.

Wall Street Is a Scary, Volatile Place

One of the scarier events that occurred in the fall of 2019 was the Federal Reserve bailing out the repo market. It was barely mentioned in the news, and you've probably never heard about it. Repo is short for repurchase agreement. The repo market is a way for banks to stay liquid and solvent by basically making short-term (overnight) loans to dealers in government securities. The way it works is that a dealer sells government securities to an investor and then buys them back the following day for a little bit more than they sold them for. This overnight market is a way for banks to move money around to fill gaps and balance their books. They basically borrow from each other overnight. It's a very fluid market. In September 2019, a situation occurred where more money went out than came back in, and the Federal Reserve had to pump more (imaginary) money into the repo market to kind of bail it out. Again, more national debt.

We've also had an overvalued stock market for a long, long time. Companies have been buying their stocks back, which inflated the market and made everybody think that everything's just hunky dory because the stock market's been going up for 10 years straight, which has never happened in history, by the way. But the fact that they keep inflating the market to make it look good and to keep everybody happy is really just kicking the can down the road.

What happened after the coronavirus is that when the Federal Reserve created this new money for the stimulus package, they didn't just hand it directly to every American. They also used some of the new money (which is now part of the national debt) to buy bonds from companies in financial trouble. In some cases, they purchased actual exchange traded funds (EFTs) on the market. For the first time, the Federal Reserve created money and then was able to directly influence private companies. They did this by circumventing the Federal Reserve Act of 1913 and applying a very liberal interpretation of the CARES Act.[3] Okay, that is scary.

Companies aren't stupid. They're like, sweet! Free money! Then they used that money to buy up all of their own stock back off the market, which made it look like there was demand for stocks and caused the stock market to go back up. This is why the market went back up after that small correction we saw right when the coronavirus happened.

Now with all this chaos going on and the markets appearing to be recovering, any real financial journalist is sounding the alarm because what's going on with the market right now is crazy. This

3 https://bit.ly/Fed-buys-corporate-bonds

has never happened in the history of the world. All the charts and indicators are going in the exact opposite directions that they should be going in order to result in the type of stock market we have right now. I don't know what's going to happen. This might blow over and become just another blip on that goofy little chart that financial planners are told to use, and that I used myself in 2008, to explain to potential investors that the stock market always rebounds. They'll be out there a year or two from now saying the same thing while the market is higher than it's ever been. That's fine, if you are 30 years from retirement, and you're willing to ride those stressful waves. But the reality is that the DALBAR study[4] that comes out every year always shows that the average investor underperforms the actual mutual funds they're in because as much as we say to buy low and sell high, people just don't do it. Emotion always beats out logic. When the market goes down, and prices get low, most people don't stay in it and buy low, they pull all their money out.

The stock market is unpredictable. What has happened in the past does not predict what will happen in the future any more than the fact that you've rolled six doubles in a row at dice means you're going to roll doubles the next time that you roll. The stock market is a gamble, no matter what your financial advisor tells you or what fancy chart they show you. It's gambling. Period.

Now, smart gamblers know how to make money, and there are certain people with lots of money who can move markets in a way that makes them money. Are you one of those people? I'm not. I don't have nearly enough money to move markets, so I'm not going to put my money in a place where I don't have any control over it, and it's at risk of just going up or down. The main characteristic of the stock market is volatility, when what we're after is financial control.

4 https://www.dalbar.com/

What Does It All Mean for You?

So that's where we're at right now as a country. What do we do about it? What do *you* do about it?

The mentors I follow say, here's how you make money. Rule number one is don't lose any money. How do people typically lose money? They lose money when the stock market crashes, and they lose money to taxes. Rule number two is to invest in things that you understand. What I like to say is that there is nothing you can understand better than yourself. Understanding yourself and the potential of your own human capital for creating opportunity is the best investment you can make. Investing in your education, your skill sets, mastering your craft, learning how to play in the digital marketplace so you can share your craft with those who need it, monetizing your craft, making profits with it that you can reinvest into other good things to help more people—that's a real investment. This is a point I have already made at great length in this book and something that I would hope that anyone reading this book is striving for. Even if you don't have an idea for your own business, there are ways to participate in the modern economy in the way that I've been talking about so that you can take control of your investment and understand it.

So many people don't understand what they're doing with their money. Maybe they decide I'm going to do 20% C fund and 10% L Fund, and put the rest in the G fund. But do you understand what you are doing? Do you understand where your money is going? Do you understand what's happening in the stock market? Do you even understand everything that I've been talking about in this chapter? Frankly, my own understanding is barely scratching the surface. But I understand enough to know that I don't want to put my money directly at risk in the market, and I don't want to recommend that other folks do that, either, because along with our lack of understanding, there is fear: fear of the market crash, fear of losing our money. We want to protect our money, which makes us tend to pull it out when things aren't going well. But then again, the government just keeps propping the market up, right? So what if we miss the upside? Now we have the fear of missing out. Well, you know, Americans don't like that.

So, is there anything out there that offers good potential gains without gambling and without fear, and that doesn't require you to go out and create a business out of thin air while you still have a good job or a career in the military? You're making good money in the military. Whether or not you are currently doing it, theoretically you could save some. What on earth can you do to protect yourself against rising taxes, protect yourself from losing money, and protect yourself from the fear of missing out?

It's not enough to just put all your money in a checking account, because then you're not beating inflation. Money sitting in an account that's not bearing interest, or that's not bearing enough interest, isn't going to beat inflation. These are real, real problems that none of the traditional financial planning CFPs are talking about. Their standard answer is that things will just "work out" over the next 30 years. Okay, but what if that year when you need *yoiur* money is the year when tax rates spike or the market crashes? I've seen many retirement accounts get depleted quite quickly when this happens.

The key to avoiding this situation is to make sure that your money is 1) not at risk in the market, and 2) not taxable.

There Are Different Types of Money

It's important to understand that there are different types of money. There's money that is taxable and money that is not taxable.

On the taxable side of things, you have your salary, or money that you earn from self-employment or profits from your business. A pension is taxed like a salary. Most people don't realize that withdrawals from traditional qualified retirement accounts are also taxed as income when you take your money out. So if you're getting a military pension, and all your money's in traditional retirement savings vehicles, that's all taxable money, and you can't control the fact that it's taxable. If you're thinking that you'll outsmart the system and just wait to start withdrawing your money until taxes go back down again (if you have that kind of time), that's not an option either. Once you reach the age of 70 ½, you have to start taking what is called required minimum distributions from your retirement account. Why? Because that money is taxable, of course, and as I've already explained, the government needs that tax money. Not only that, if you want to pull any money out of your retirement savings accounts before age 59 ½, you have to pay an extra "tax" to do that, a penalty fee of 10%. Why? Because if you had left your money in there for longer, it may have grown substantially greater, thus yielding a much greater chunk of tax income for the government, and the government doesn't want to lose potential income. You can't control this either.

But say that you wait until after you're 59 ½ to start accessing your traditional retirement savings and avoid that penalty fee. Your military pension is taxed at a certain rate, but as you start pulling additional money out of your retirement account, this additional income increases your marginal tax rate, which means that the money from your retirement account is taxed at a higher rate than your pension is. So if you need $20,000 to live on in addition to your pension, at today's tax rates you may have to pull out close to $30,000 or maybe even $40,000 in order to end up with that additional 20,000 after taxes, so your account gets depleted faster, and that's at today's low tax rates. Because of the whole tax situation, it's important to understand that the amount of money you have in your retirement account is NOT the amount of money that will eventually end up in your bank account.

Another kind of taxable money is called capital gains. If you have some traditional investment like real estate, or something that's growing in value like stocks, bonds, or mutual funds, the money that those investments earn is called a capital gain, and you have to pay taxes on any capital gains. So as your stocks, bonds, or mutual funds pay you out a dividend (a sum of money typically paid quarterly by a company to its shareholders out of its profits), you have to pay tax on that money as well.

So where on earth can you put your money where it's not taxable?

There are certain other financial vehicles you can put your money into where it will not be taxable. For example, municipal bonds are tax-free. A municipal bond is a bond issued by a local government or territory (or one of their agencies). The money generated by the purchase of municipal bonds is usually used by the issuing entity to fund public projects like building or maintaining roads, schools, airports, and other infrastructure-related repairs. A lot of rich people park millions of dollars at a time in municipal bonds. These bonds only pay like 2% interest, but 2% on a million dollars is $20,000 of income, not too bad for money you have no immediate use for yourself anyway, right? These vehicles aren't real sexy, and they're not really easy to use, but they are a source of tax-free income for the very wealthy.

Another tax-free vehicle is the Roth IRA, Roth 401(k), and Roth TSP. You fund these vehicles with money that you have already paid taxes on, so you don't have to pay taxes on any money you pull back out of them. This not only goes for the money you contribute to a Roth, it also goes for whatever growth you see on that money. This seems like a pretty good deal, as it allows you to avoid one of the traps we are trying to avoid: future taxes. But there's a catch. You can't really put much into a Roth IRA account. There is a contribution limit every year, and from time to time they change that limit. Right now it's around $6,000. When you're in your 50s, you can also do a "make up" contribution of an extra $1,000 a year. Whoo boy. But notice that the government wants to limit how much money you can put into a tax-free account. Not only that, once you make a decent amount of money every year, another cap kicks in with regards to how much you can contribute to a Roth plan. If you earn over $200K a year, you can't use it at all.

So, municipal bonds have a low rate of return and tie up your money for years at a time, and because of the limits on Roth accounts, the Roth can only take you so far, and only give you half the protection that you need anyway. But there is a third place where you can put your money where it is both protected from market volatility and future taxes, and I'm going to talk about that third option in the final chapters of this book.

What Can You Control?

This book is about how to achieve autonomy, and as I've shown so far throughout this book, autonomy is really not possible unless you can control what is going on around you. Obviously you can't control every single thing that happens, but you can control whether or not you will have the resources that you need to deal with the things that you can't control. The question is, given the overall state of our economy today, what can you control? What is it possible for you to control?

We can control how we invest our money. We can choose to invest in ourselves. I have dedicated a big chunk of this book to talking about how to invest in yourself. In the remaining chapters of this book, I will be talking about how to control how you invest in others. We can control how much we pay in taxes today by placing our money into tax-deferred vehicles. We can also control how our money reacts to market volatility based on what kind of vehicles we put our money into, and for the remainder of this book, I am going to tell you how to do exactly that. As I have said, the rich know how to position their money in a way that insulates them from these kinds of problems. They use specific financial vehicles that operate according to little-known rules that protect their money both for them and for their future generations. Decades ago, these financial vehicles were only available to the very rich because the price of entry was far beyond what most people could afford. Today, they are available to anyone and everyone. Today, they are available to *you*.

17

Pillar #3: Introducing the Secret

For most people, the fear of loss is much greater than the desire for gain.
~ Anthony Robbins

There is an asset class with more flexibility and stability than any other asset class available to you at this time, and I can teach you how to use it. This asset class:

- Allows your money to grow based on the performance of the stock market without actually putting your money at risk in the stock market;
- Grows your money tax-deferred;
- Allows you to access your money at any time, for any reason, without having to pay tax;
- Protects your money against litigation;
- Can be used to pay for medical costs or long-term care costs;
- Can be used as an income stream;
- Allows you to create a private lending solution where you act as the bank, either for yourself or to create a business to lend people (like real estate investors) money;
- Protects your estate from estate taxes so that more of your money goes to your family;
- Can be used to offset any costs that you would normally pay for a similar service. For example, it provides the

same benefits as the SBP for significantly less than the SBP would cost you for 30 years.

This asset class has been around for 150 years, and new technology has made it more accessible than ever. As someone in the financial industry, I have learned about lots of different financial products and tools I can use or recommend. But if you want financial control, then this asset class is the one that gives you a foundation from which you can leverage your money to create opportunity for yourself. Opportunity usually means more income.

What is this asset class?

It's not bitcoin. It's not gold. It's not tax liens or municipal bonds. It's an investment grade, private placement, free market, life insurance policy called Uncapped-Indexed Universal Life. If your father's life insurance policy was like an old station wagon, this is a Tesla.

In the last three chapters, I went down into the weeds about compound interest, assets, and taxes because these concepts are fundamental to being able to make informed decisions about what you want to do with your money. I was also laying the foundation for this statement that I made earlier, but which is so important that I am going to repeat it:

You have achieved financial autonomy when you have produced assets and income streams that do not require your constant attention and that do not require you to gamble.

I hope that you can now see why the traditional financial vehicles that we have discussed do nothing to further your journey into autonomy. The financial vehicle I am about to discuss, on the other hand, *will* give you autonomy.

The third pillar of autonomy is financial control. The truly wealthy have assets, and they leverage them to create income streams. They have systems set up that make money no matter what they are doing, even while they're sleeping! These are awesome, super important things, but equally important, maybe even more important, is that the wealthy understand how to protect what they have. The 1% in this country are not immune from financial devastation because they have so much money that such a thing isn't possible; their fortunes are able to remain intact even when stock markets crash and the economy tanks because they know how to protect them. But before I can effectively tell you what tools they are using, I have to blow away yet another common myth that you have probably always believed.

What You Probably Think Life Insurance Is

What's the first thing that you think about when you hear the phrase "life insurance?"

Most people think of death.

Why? Because that's what we think life insurance is about! When you get a corporate job, life insurance is commonly offered as a benefit. This is because the life insurance that comes along with a corporate job is only about your death. The lump sum paid out to a beneficiary is a benefit – for them, not for you. When you insure you own life as an income-producing asset, however, much as you protect your other assets like your home or your car against loss, life insurance becomes an asset.

That life insurance is a death-related benefit is just one more of the myths and misconceptions that so many people live by, and that make it so much harder to get their financial lives in order than it ought to be. I am going to talk about life insurance in the context of life. I want to dispel this myth and finally wake people up to the realities of our financial industry so that they can take advantage of this knowledge for themselves.

Insuring the Income-Producing Potential of a Life

Pretty much everyone thinks that car insurance is for accidents, homeowners insurance is for fires (and other catastrophic events), and life insurance is for death.

Well, that's just incorrect. Car insurance and homeowners insurance are products. A product represents a net financial loss. You will probably never see any return on the money that you spend on these insurance products. Even if/when you are unfortunate enough to have a major car accident or a house fire, over the lifetime of these products, you will not recoup as much money as you spend on them. That is simply a fact. After all, insurance companies are businesses, and businesses operate in order to make a profit. If most of their customers ultimately received back more money than they paid into their plans then none of these companies could stay in business.

Term life insurance is also a product, because you only pay into it for a specified period of time, ten, twenty, or thirty years, during which you are statistically highly unlikely to die. When the agreed-upon term ends, so does your coverage, and you are left with nothing but the loss of the twenty-five, thirty, or forty dollars a month that you spent on that product over the term. Permanent life insurance, on the other hand, is not a product. It is an asset class. Real life insurance is about insuring the potential of a life, and it can be used to the benefit of the person whose life it insures. Permanent life insurance offers a death benefit, but it is NOT about securing a lump sum at the end of the insured's life.

Term life insurance policies are typically sold to customers based on fear of missing out. Poorly trained insurance agents and financial advisors – and remember, these are not people who necessarily understand finances, these are salesmen – learn scripts designed to convince people to purchase death insurance out of fear. "What if you died tomorrow Mr. Johnson? Your family would be destitute."

No shit. That's not helpful, and it's not motivating someone to see opportunity.

That's why I wrote this book, to *communicate* that what you have probably been led to believe about various asset classes has only been part of the story. Because financial advisors and realtors and insurance salesmen are constantly failing out of the industry, it is always full of new guys. Of course they aren't in the industry long enough to truly become experts on financial products, and anyway, their firms don't want them to be experts, they just want them to sell! Consequently, how much can you really rely on any "advice" that these so-called "financial professionals" give you?

What It Means to Insure Your Life

Veterans coming out of the military have options. They can get a job. Or maybe they don't want a traditional job; maybe they want their independence and freedom, which means they want money and time. I think that those who recognize the opportunity right now will see that in the long run, getting a traditional job is the more difficult route. When your primary focus is to get your time back, then you start thinking about what you're going to do with that time that you want so desperately. For a lot of people, the process of figuring out how you are going to make that happen allows creativity to come out. This is because they inevitably find that they want to spend that time providing value, and that means creating.

That's what the human species is here for. Some people create things directly, and some are more creative at doing the hard work that's needed to help other creators get their work out into the world. But we are *all* creators, and the difference between the traditional job and the entrepreneurial path in most cases is the difference between working to realize someone else's vision and creating your own.

Once you start to create, you're producing something. You're giving back to society, whether you're making a widget, painting, or whatever you're doing. You are an income-producing asset, and as such, you have a financial worth. How much money do you make in a year? How much money do you make in ten years? What if you invest a portion of your income every month? If you were no longer able to make that monthly investment, what's the potential for loss there?

You probably think that I'm asking these questions within the context that you might die suddenly and unexpectedly, but that's not why I'm asking at all. I'm asking because what if you decide you don't want to work anymore? Or what if you decide you want to change fields and start a new career? Or travel the world? I'm talking about insuring yourself as an income-producing asset so that *you* can reap the benefits of the insurance policy (i.e., access the money you have put into it and which has subsequently grown) while you're still alive.

So many people think that they are stuck in jobs that they hate. And they are, in a sense, because if they quit their job, they are cutting off their income. They haven't insured their income in a way that can benefit them. But if we get our priorities right, then we can have so much more control over our lives. Insuring the income-producing asset, you, is mobile, cheaper, and easier than other kinds of investments like the stock market or the real estate market. It's no different from a retirement plan or a real estate strategy in that it's simply another financial vehicle that you can leverage to build your portfolio.

Are You Overlooking an Investment in *You*?

If you're not thinking about insurance as an asset class then you're missing a whole aspect of your financial life. You're just completely ignoring it. It's kind of ridiculous that we've taught everybody to do that. I'm thinking it's because of all the shysters in the 80s. You know they existed, but so did stockbrokers and so did the whole subprime mortgage real estate bubble that was just over a decade ago.

I think insurance has been ignored because the insurance companies were being smart and innovative from places like Des Moines, Iowa. They weren't being flashy. They weren't running Super Bowl ads. They weren't trying to beat the stock market.

They didn't have any Bernie Madoff situations. They didn't have any Lehman Brothers situations. Heck, they didn't even have any Mom and Pop credit union FDIC takeover situations. They were just sitting nice and quiet.

They're not good at marketing themselves. Frankly, they need people like me to recognize their value and apply it in such a way that we can get this asset class into the hands of more of the right people. Finding the right people is key, just like everybody shouldn't be using the VA loan and buying real estate. The VA loan is not an asset. It's the privilege of not having to pay a lot of costs that other Americans have to pay when buying real estate. It's a perceived benefit, but that doesn't mean you have to use it. I think it's a crime that people are sold VA loans by being told that you've earned it, and therefore you should use it right away.

Maybe. Maybe not.

Reputable financial advisors, real fiduciaries, should be giving their clients better long-term advice about how to create wealth instead of just hoping the stock market works out in their favor. They should be educating all the new butter bars out there who have an extra thousand dollars a month and who want to go buy a Ford Mustang about the other, better things they could be doing with that money, things that will increase its value rather than draining value out of it because there's going to come a time when they are going to need that money. Look at it this way. Military life is about following orders. All kinds of things might cross your mind that you might like to do, but they are all inevitably met with a big *no* while we are in the military. So the one thing we have to look forward to is transition, and when we approach transition with fear because we don't know what to do with ourselves, we're in the completely wrong mindset. If we haven't prepared for it at every level?emotionally, spiritually, socially, and financially?then at transition we hear a whole lot of *no*s all over again. At least we think we do, and we say a whole lot of *nos*, too. Not because we necessarily want to say no, but because we haven't set ourselves up to have options.

At that point, transition becomes all about money. It retroactively redefines our service to our country as having been just a job, when we signed up to serve. Let's stop talking about the VA loan like it's a handout. I'd say it's a bonus, and it's usually a good idea to try to do something smart with a bonus. We need to respect the tax-payer who is paying for these loans and for our pensions and VA disability benefits by being good stewards of that money.

In Chapter 14 I talked about compound interest. Most of us have been on the wrong side of compound interest. We just kind of ignore the fact that we're not truly benefiting from it, and the opportunity cost of that negligence is like a tax. Overpaying for some things is also a tax. We're all overpaying for our insurance in the military, since for the most part, most people are healthy. Those dollars that are taken out of your paycheck for insurance every month are kind of redirected in a weird way, but they essentially go straight into the pockets of a government contractor. Why not privatize that stuff? Why not prepare our service-members to be veterans by giving them the freedom to choose and use their own tools and resources instead of just teaching them a cookie-cutter approach to financial planning?

A veteran is no longer in the military. We have assets as a result of the time that we sacrificed for our country, but we don't have to follow orders anymore, nor should we. The thesis of this book is to show veterans who have a very, very opportunity-focused mindset how to leverage the new financial industry technology that's been made available over the last two decades and perfected in the last few years, specifically on the insurance side, to their benefit. I recognize how it's creating opportunities for veterans and military spouses to live a more intentional life, and I want to tell you about it.

A Quick Recap: The Financial Realities of Apathy

The belief is that the only way to save for your future is to use a convoluted mess of financial vehicles, mostly geared towards retirement, in order to beat inflation and amass enough money so that you can feel confident you'll be able to stop working one day and live off of that savings. Maybe you've tried everything from stocks, bonds, mutual funds, IRAs, bank accounts, CDs, or even real estate to try to get growth on your money. You also want to protect it from risk, and the tax code is so freakin' complicated that nobody really truly understands it. This perfect storm of lack of financial education and understanding is why everyone feels forced to hire financial planners, tax attorneys, or CPAs. It allows the financial planning industry to justify their need for you to pay them to find the perfect mix and match of these all these different financial instruments that are really confusing to understand and hard to manage.

If you don't understand finances, and you don't think you can understand it, I get it. I've been there. We don't learn about how money works, if we ever do at all, until we're out of school and we get our first job. For most of us, our first introduction to the realities of money are when we get our first paycheck, and it's a hell of a lot lower than we expected it to be due to federal and state taxes, FICA, and social security withholdings. If you learn about investments, it's probably because a financial planner gives you a sales pitch and explains all the complexities. Few of us really take much more away from those sales pitches than, "Man, I didn't know that. I guess I need to do all these things."

I got that pitch as a young cadet getting out of West Point. We were easy sells. I bought mutual funds and a Roth IRA through the guy that pitched me and just left them there for years and years and never paid any attention to it. I started my TSP when I was in Iraq and started saving money, and I'm glad I did. I even had an extra $10,000 in my checking account I wasn't expecting to have when we got back from Iraq, and I remember asking my dad what I should do with it. I was 25 at the time. He recommended a certificate of deposit, so I went and got a CD. I didn't need a CD as a 25-year-old, but my dad was 63 at the time, and that's what made sense to him. We get sold this story about compound interest and how important it is to get compound interest so that our money can grow enough to beat inflation. Inflation is just the government devaluing the currency. It's another tax that justifies the need for everybody put all their money at risk in the stock market in order to keep everything going. It's a big wheel.

We need to shift the perspective on this. That's why I wrote this book.

First off, the cornerstone of most financial planning is retirement. Nobody ever bothers to ask you if you even want to retire in the traditional sense. Sure, if you have spent 40 years doing manual labor, yeah, of course you want to retire from that, but fewer and fewer people these days have those kinds of jobs. The world is different now. If you only save for retirement in a traditional retirement plan, then you are locking all of your money and resources into a system that is hard to manage, is very risky, that costs a lot, and that limits your access to your own money until most of your life has already passed you by, all for a goal that really is probably pretty darn low on your priority list anyway.

Most Americans aren't really saving for retirement. I'm not saying that they shouldn't be doing it, I'm saying that they're not doing it. Instead, they're working for the weekend. They have too many bills and expenses to pay now to think about savings. The national debt has gone crazy, and most Americans have gone crazy with it. Take the recent coronavirus stimulus checks. Most people used it to go buy something, and that's a missed opportunity. People put off saving because it's hard, and there is no immediate or even short-term reward for doing it. This is why it's so hard to get clients when you're a financial advisor. I used to think it was because there is no shiny object a financial advisor can use to captivate people into spending their money. When I was working as a financial advisor, my office was inside a building with a BMW dealership in it. People would come in to buy the BMWs, but they wouldn't come in to save money for their future. I was right upstairs. I could have helped them buy the BMW more efficiently, but no one came into my office. They came in for the shiny thing, and avoided me because they didn't want to be told no.

Later, I realized that it's not that there's no shiny object, it's that there's no motivation. When you're 20, 30, and even 40 years old, it's not motivating to save for retirement. What 30-year-old thinks about how to set themselves up financially so that they can stop working at age 65? I sure didn't. It wasn't exciting to me. I didn't care about retirement. I still don't. I'd rather keep working. I mean, I choose to live a lifestyle of autonomy, where I can travel and be anywhere I want and have mobile income. Retirement in the traditional sense is irrelevant to me, and in this day and age, it could be irrelevant for you, too. Saving for traditional retirement means that you are selling your own personal time to make money, handing that money off to other people to manage, and hoping that they grow it so that you can use it to buy back your personal time much later in life, at a time when you will probably be faced with health issues.

In the next chapter, I am going to show you that there is actually a pretty simple way to handle all of these things: savings, getting growth on your money, taxes, retirement planning, and being able to live the life of your dreams sooner rather than later.

Do you know how many career colonels who are approaching age 50 know they haven't taken saving for retirement seriously and are worried that they need to start? Why do you think that is when their take home pay is over $160K?? It's because they've been using all their money to take care of their family and live a good life, but now the kids are going off to college, and they need to play catch up. The traditional financial advisor will just say, "Whelp, guess you need to save more and work a lot longer. Here's your portfolio. That boring advice will cost you 1.5% of your investment compounded annually." Or they could do a strategy that I teach which leverages the participating loan provision and tax benefits and be done in ten years, simply by repositioning their existing assets, including future pension payments. This is a strategy that offers a shiny object rather than the uncertainty of having to work and save longer.

You can revolutionize your finances with a tool that insures the income-producing potential of you as an individual. This tool insures you against losing your assets to the government in the form of taxes. It insures you against the tax of inflation. It insures you against the risk of outliving your savings. It insures you against the inevitable high costs of healthcare. It insures you against needing to go to the bank to borrow other people's money to get what you need. It gives you resources that allow you to create investment opportunities, whether that means starting a business or being a private lender. It allows you to create a private pension for yourself so that you can stop working for other people and live off of that pension whenever you want.

It's not some new tool like Bitcoin, although when the cryptocurrency version of this financial tool appears, I'll be the first to learn it. It's actually a pretty standard financial vehicle, it's just that no one else has seen the motivation factor that's possible by not making saving all about retirement. That's where I'm different. Yes, this same tool that I am about to tell you about gets pitched to people as an alternative retirement strategy. But it can also be used to create the "shiny object" that you want in the nearer term – three, five, ten years from now, rather than 20, 30, or 40. It can make possible whatever it is that motivates you *now*.

18

The Financial Control Secret Revealed

Learn the rules like a pro, so you can break them like an artist.
~ Pablo Picasso

This financial tool is called Uncapped-Indexed Universal Life (IUL). It is an advanced equity life insurance strategy. Go ahead and Google it right now. Let's get that nonsense out of the way.

The 10 or maybe even top 50 entries are all probably going to say something negative about an older version of this, because the people who wrote those reviews are talking about a different iteration of the vehicle, and they don't understand that there's been innovation. This makes sense when you consider how many people have had products like this sold to them in the wrong way (by salesmen who didn't understand them either), or how many have misused it and lost money because they bought something that they didn't understand. In that respect, an IUL is no different from anything else. If you buy it for the wrong reasons, or you buy it without really understanding what you're buying, you're going to be disappointed. Think of it like buying a car. If you buy a stick shift thinking you're buying an automatic, and you've never driven a stick shift, you're not going to know how to drive your car, and you could damage the car trying. Let me say this again, because it is so important:

Buy things that you understand.

I want you to understand the IUL, so I'm going to teach you how it works. I'm going to explain all the features of this vehicle, how much it costs, exactly where those costs are, and how those costs diminish and even turn into income over time. Then, I'm going to walk you through a scenario with someone who is 40 years old, who has been in the military for a few years, who is planning on serving his full commitment, and who is probably going to finish out their 20 years in the military. They know they're going to get a pension. This person has been doing the right things. They have saved a bunch of money in a Roth IRA and in their TSP. They have saved up a nice emergency fund, and they have a big term life insurance policy in place to protect their family in the event of their premature death. Guess what? There is nothing wrong with any of that whatsoever. The point of this book isn't to discourage anyone from doing any of these things, and if you have already done these things, then congrats to you because you are already in the top 10% of Americans when it comes to financial stability. The vast majority of your peers aren't doing these things. So whenever I see someone who has managed to put themselves in that situation, I applaud them because they have really set themselves up to take advantage of what's going on in this economy.

That said, they've also planned and saved themselves right into a trap, and I want to be sure you're aware of it. The trap is that traditional financial planning theory is designed for the average American with an average level of income. Here's the shocking reality: once you get out of the military, start collecting your pension and VA benefits, and get additional income from a new job or business, then you are quickly vaulted into the top 5% of income-earners in the United States. In fact, if your family makes just $250K annually, you are already in the top 1.5%![1] And guess what? The top 1.5% of people don't typically do their financial planning like the other 98.5%!

1 https://bit.ly/affluence-in-United-States

Maybe you aren't there yet. I'm certainly not. I don't have a pension, so I'm only relying on my business. But I'm patient because I see the potential and would rather spend 4 or 5 years eating shit learning to do something that I love to do instead of spending 40 years in a "safe job" doing something I hate.

The financial strategy and examples I'm going to teach you in the rest of this book are really only scratching the surface. You'll be able to see plenty of other use cases and ideas on my YouTube channel. Nor is this the one size fits all approach that would magically solve everyone's problems. Instead, think of this as the new foundation to wealth. The one that simplifies and even removes the need to waste time, costs, and energy dealing with the anxiety of traditional retirement planning complexities, especially if you're someone who strives to make more than $250K per year, someone who strives to be in the 1%.

Ready to get into the weeds on the IUL? Okay, here we go.

Before I begin, I want to say once again for the record that the traditional financial vehicles available to veterans are not bad, especially for those who are crystal clear on their financial goals and for whom traditional solutions align with those financial goals. The problem with traditional financial vehicles is that they are rarely managed properly, and they don't give you much flexibility. For those open to other options, an Uncapped-IUL offers some distinct advantages that you will not find in these traditional financial vehicles. We help our clients use this other option to put some, not necessarily all, of their money into a secure asset that beats inflation. With no volatility, no risk of market losses, and no requirement to pay taxes on income, they can be much more confident with their assets. Veterans then have a safe, secure vehicle from which to leverage their money for post-military life opportunities, decades before they reach traditional retirement age. It's a no-brainer option for a certain kind of individual that still allows the entrepreneurial-minded veteran to save for retirement. Only you can make the determination whether or not you can benefit from an IUL.

19

The Secret Explained

A creative man is motivated by the desire to achieve, not by the desire to beat others.
~ Ayn Rand

In the banking world in 2008, the financial crisis was based on real estate and over-leveraged things like credit default swaps, and everything seemed super scary. Major banks went under. Hundreds of small banks had to be raided by the FDIC. Stock values fluctuated. Real estate markets crashed. Yet, like always, the insurance industry just kept truckin' along. Only a handful of insurance companies have ever had problems in decades. Part of this stability is thanks to the industry's forward thinkers that continue to innovate in how they approach managing assets for their policyholders.

Innovation in insurance is just like innovation in every other aspect of our lives. It's a new technology, and people haven't been able to figure out how to employ it, especially the financial professionals, because it's a "bringing your new toys to an old fight" type of approach. But I believe that within twenty years we are going to see a lot more people protecting their money with IULs because of the way the financial system is going to play out and the amount of true risks around money—whether it's the risk of a stock market crash, the risk of taxes going up, the risk of getting sued, the risk of losing jobs, or the risk of just plain not having liquid assets in the right places. As with innovation in any industry, there are growing pains. Permanent life insurance was designed for the wealthy, so bringing it to the middle class has been messy, which accounts for some of the bad rap the IUL has gotten from the old-school financial advisors. But with new advances in technology and methodology, and with a lot of trial and error, I can certainly say that it is now an investment option available to the middle class.

The IUL offers the following unique advantages, and I am going to talk about them each in turn in the sections below.

- An IUL allows you flexibility in funding. You can contribute as much or as little as you want, over as short or as long a period as you want. You can also fund the policy with a single lump sum spread out over a few years.
- An IUL allows you to invest money in the stock market and benefit from that growth without any risk of losses when the market declines.
- Money you invest in an IUL is contributed post-tax. This means that not only do you get the advantage of paying taxes at today's lower rates, you also won't have to pay taxes on any growth when you withdraw it. Unlike a traditional retirement plan, with which you are guaranteed to lose 10%, 20%, 30%, or who knows how much in value when you withdraw your money in the future as tax rates rise, the money that you have invested in an IUL is the amount of money that will eventually end up in your bank account when you are ready to withdraw it.
- An IUL allows you access to "your" money without actually withdrawing your money so that you can still benefit from the market growth on that money. What this means is that you can use your money for what you need while it continues to grow in value when the market is on the rise. I will explain how this works below.

- An IUL allows you to access your money at any time, for any reason, without penalty. You do not have to wait until you reach retirement age to access your money and growth. You can also use your money as an income stream for retirement, just as you would a pension or a traditional retirement account.
- Money you withdraw from an IUL is tax-free, so the cash surrender value of the account is the amount of money that will eventually end up in your bank account when you are ready to access it.
- An IUL offers a death benefit, just like any life insurance product, without being about the death benefit.
- By setting up your policy with the lowest possible death benefit, we can keep your costs lower, and reduce the amount of time it takes to fully fund your policy and thus the amount of time before you can begin to access your money.
- An IUL offers additional benefits like the ability to pay for long-term care for yourself or a family member, litigation protection, and the ability to leave a legacy to the next generation.

Does It Cost More Initially? Sort of… Not Really

An IUL is different from other assets because it becomes an asset which you can control. This allows you to build equity and have flexibility. You can determine the stipulations of the contract, how you fund it, for how long, and for what reason it can be used throughout your life. Think of it like owning a property. The property doesn't have to be a single family home. You might turn it into a duplex, make it into a lawyer's office, or tear the whole thing down and build something else on the land. Unlike with a term insurance policy or a TSP, which have rigid stipulations and a single purpose, with an IUL you have options.

An IUL is also comparable to an investment in real estate in that it requires more investment capital. You have to put in more money up front to get into an advanced contract with a highly reputable financial institution, but you can maintain and even control those costs over time because you're building equity and buying down the amount of insurance. This is similar to buying a home. When you finance a real estate transaction you pay off the bulk of the interest on the loan in the earliest years that you are making payments. The longer you make payments, the more of each payment is going towards paying down the principal rather than servicing the interest. Also like buying real estate, an IUL policy will be there for you in the end, as long as the asset is used properly. As long as you pay the mortgage, you will eventually pay off the loan, and the house is going to be yours to pass on however you want.

Because this is life insurance, as with a term policy, the insured (you, a spouse, a child, etc.) has to qualify and be healthy so that you or the insured on the policy are a decent risk for the insurance company. But don't worry if you have health issues. We're not using this strategy because of the death benefit. That's just a bonus. We want to take the offensive approach on the healthiest insurable individual (it doesn't have to be you) to get the best return on investment.

Apart from that similarity, term life insurance and IULs are such vastly different products, it's like comparing apples and oranges. The only other similarity between them is the fact that in both cases the policyholder is being underwritten for a certain amount of death benefit should a catastrophic event occur. With term insurance, death is the only time there is a benefit. But with an IUL, the policyholder is leveraging it for the inevitable (of course we're all going to die), but they are also getting tremendous tax benefits, asset protection, legacy and wealth transfer benefits, and so much more than just sinking money down the drain for a term policy that in all likelihood will never provide a return on your investment for the beneficiary, and certainly never will for *you*.

So yes, one of the main criticisms of the IUL is true. It requires more money up front, and that's why it's easy to complain about it being expensive compared to term insurance and from the point of view of the mainstream media and standard financial planning 101 stuff. But the reality about these costs is that the IUL costs less than most things over the long term. The commentary about high "costs" are more around the fact that the IUL works better when funded aggressively.

The key to the financial advantage here is time and liquidity of assets. In the short term, an IUL requires significantly more commitment than a traditional retirement plan. This is a tool meant to be used, not managed like a 401(k) that you haven't looked at in years. However, once the IUL is set up, provided you commit to it and hang onto it, after ten years the costs drop to almost zero, and as I explained, you begin to recoup some of those costs in the form of bonus interest credits. A traditional retirement plan works in just the opposite way. In the early years of the plan, when your balance is low, your costs are fairly low. But as your balance grows, so do your costs. As you can see in the table below, at around year 20, the IUL becomes and remains the much better deal, cost-wise.

There are far too many factors around long-term investing and saving to make cost the most significant thing to compare, but since everyone is so focused on it, here's what that cost comparison looks like (below). But remember what's being paid for. The fees for investment accounts are considered the cost of money management, along with any administrative fees. TSP and index funds are far cheaper because there's little to no management. The argument usually made is that most managed accounts don't do better than the market, so why not avoid the fees and just get the index fund? If that's all you're looking at, and you're willing to take the risks that responsibility offers, then in that context, that choice makes sense. But most people have the majority of their money in accounts they pay management fees for. They don't even think about it, let alone try to justify it to themselves. It's just another tax they pay for something that they think is being taken care of for them.

A quick comparison shows why comparing costs of insurance to other investments is really a joke. In this example, a 40-year-old male makes a $10,000 annual investment for 10 years for a total of $100K invested and growing at 5.5%. The fact that with an IUL you get bonus interest credits to repay costs is a game-changer. By year 30, not only have you gotten all your costs back, you've made money! And that's not even including the growth.

Financial Vehicle	Cost at Inception	Cost at 10 Years	Cost at 20 Years (and bonus interest credits)	Cost at 30 Years (and bonus interest credits)
IUL	$1,251	$14,456	$16,761 – $8,375 rebate = $8,386	$20,224 – $24,886 rebate = $4,662
401(k) (1.5%)	$150	$9,585.13	$32,327	$65,764

The IUL is not for everyone. It is for those aiming to be in the wealthiest 1%. Following Pareto's principle or the 80/20 rule, we can expect that 80% of people are going to be doing the basic status quo stuff, and if that approach provides you with what you need and want in life, and you are happy, then hey, great. Keep doing what you're doing. But there are veterans out there who are relatively financially stable and who understand that "Hey, I'm making a good paycheck in the military, and I can easily put 10%, 20%, or 30% of my income into savings. I don't want to just hope that the stock market works out and that my money grows. I don't want to wait 30 or 40 years before I can use it, either. I want to understand what's going on with my money. I want to protect it, and I want it to be there for me whenever and however I need or choose to use it." These folks are part of the 20% of the general population who are doing things differently. These are the veterans who want to position themselves financially to continue to serve their country in post-military life.

The Costs Explained

In most things, there is a relationship between cost and efficiency. Inefficiency can increase costs, and efficiency can reduce them. This is also true of life insurance. Life insurance can be very expensive if it's not properly designed or if you're using the old technology like Whole Life, but it can also be very cost-effective and involve lower fees if it is properly designed and administered. Over the long term, say, a 20-year window, the average cost with a life insurance policy the way that I know how to design them is less than the cost of a mutual fund. So if you've heard about gigantic costs around life insurance, then you've been subjected to fear-mongering because the reality is that properly designed and administered, an IUL is going to cost you a lot less than your traditional retirement plan.

There are three main costs associated with a life insurance policy: the premium charge, the cost of insurance, and the policy issue charges.

The Premium Charge

The premium is the amount of money that you are putting into the policy to fund it. The premium charge is about 5.5% of your premium. The actual premium charge varies slightly depending on which company issues your policy. For purposes of this example, every time you pay your premium, you pay an additional 5.5% of that premium in costs. This is a profitability center for the insurance company. This is the "skin in the game" they require from your investment for them to set this all up. They use this charge instead of charging on the balance of your assets in perpetuity like with a retirement account, and this works heavily in your favor.

When you set up a policy, you determine the length of time over which you want to fund the policy. For example, say you're going to contribute a lump sum of $10,000 a year for a period of 10 years. That 10 years is called the funding period. With each $10,000 premium payment, you also pay a $550 premium charge. If you spread that $10,000 a year out into monthly payments of $833, then each month along with your premium payment, you would also pay a $45.83 premium charge. When you reach the end of your funding period, the premium charge goes away and you never pay it again.

Policy Issue Charges

The policy issue charge pay for your underwriting. It covers the cost of delivering the policy, compensation for the agents/brokers, your medical exam, and all the little things that need to go on in the background in order to issue a policy. Again, this charge only applies during the first 10 years of the policy, and then it goes away.

The policy issue charge is calculated on the basis of a per $1,000 of your policy's specified amount of death benefit and varies according to factors such as the insured's gender, age, specified amount, and risk class. The policy issue charge is calculated when the policy is issued. The insurance company then deducts this charge from the current accumulation value for the first 10 policy years.

The policy issue charge can fluctuate depending on your age, heath, etc. For example, say that you have set up your policy to provide a $500,000 death benefit, and you're going to fund your policy with ten annual payments of $10,000 each. Say you write one $10,000 check and then slip and fall and die getting out of your bathtub. Your beneficiary will get a check for $500,000! So you paid $10,000 into the policy, and your beneficiary gets 50 times that. That's a pretty good return on investment for you, but it represents a loss for the insurance company, so they need to protect themselves against this possibility.

With these new IULs, the 2.0 version of permanent life insurance, we have the option of reducing the death benefit to reduce our costs. Whether or not and how much you want to do that will depend on your goals for the policy. If your goal is just to provide a massive death benefit for your heirs, you can do that. But if your goal is to use the policy as a cash accumulation vehicle, then there are ways to make your policy as efficient as possible towards that goal. The death benefit is really where the cost of these policies are. This is neither good nor bad, it's just a fact: a higher death benefit equates to more cost, a lower death benefit equates to lower costs.

When we design your policy, we set it up to achieve a balance between the death benefit and the cash accumulation value that is specific to your personal goals for the policy. Generally speaking, in most cases, our main focus is on optimizing the cash accumulation value. So if your goal is to be able to access more cash, then we set up a lower death benefit. If your goal is to have a higher death benefit, then you will have a lower cash accumulation value and access to less cash in the policy. We can reduce costs by thousands of dollars simply by reducing the death benefit. Again, this is an option, not a requirement. Also, the amount of the death benefit is not permanently fixed. Over time, as you are able, you can increase it as needed if you've used a convertible term policy on the defensive side of this strategy.

Cost of Insurance Charge

This charge covers the base cost of providing insurance under the policy. The amount depends on the insured's age, gender, health, underwriting class, policy experience, and death benefit amount. This charge applies throughout the life of the policy.

Additional Charges

The only additional charge that applies across-the-board is a little administrative fee of about five bucks a month, which is insignificant. Other additional charges could include transaction charges to cover the costs of preparing the annual paperwork, charges for activities (skydiving, swimming with sharks, etc.) or health conditions that exceed the company's risk standard, and costs associated with selected agreements, if applicable.

Concluding Thoughts About Costs

For most people who are still on active duty, even with something in the neighborhood of half a million dollars in death benefit, we're talking about costs of a few hundred dollars a year. That's it. Some companies, which are not owned by shareholders but by the policyholders, actually start giving you a rebate on these fees once the policy is fully funded. So in our example above, if the policy is fully funded in ten years, then in the 11th year you will start to see a "bonus" or a rebate credited to your account every year, in addition to whatever you might make in the market, which means that in this situation, over the long term, your net cost could eventually be zero.

You are probably also wondering where my paycheck fits into this picture. As a broker, the insurance company pays me up front for introducing the client to the carrier. The insurance industry has worked this way for 150 years. These companies need an agent to design each policy to fit individual clients' needs, not to mention to find the clients in the first place. These insurance companies suck at marketing, so they pay brokers very well to bring them business.

This payment doesn't come out of the upfront transaction like a real estate agent's commission, however. It's part of the policy issue charge that is spread out over the first 10 years of the policy. From year 11 on, the insurance company then issues rebates which are contractually guaranteed to repay some of the policy issue charge, because if you still own the policy at year 10, there is a 90% likelihood you will continue to own the policy with the company for a very, very long time, and they know they'll make money and you'll make money. My payment is spread out across the initial 10 years of every policy because there's got to be skin in the game to help the company offset the costs for the risk they are taking around the death benefit in early years, as well as the costs of the many layers of underwriting and the costs of marketing tools that they provide to brokers like me to be able to introduce you to this amazing tool.

Flexibility in Funding

Because the IUL is a tool used primarily by the wealthy, you may assume that you need a large chunk of change as an initial investment, but this is not the case. You have complete flexibility in funding. You can start and fund a plan at whatever level is affordable for you. You can pay smaller amounts of money over a long period of time, or larger amounts of money over a short period of time. You can add big lump sums or spread it out month after month. The only limitations are the amount of insurance you can buy based on the health qualification and IRS limitations depending on your age and the amount of insurance you want to buy, should you not want to create a taxable event. Basically, there are thousands of options and flexible ways of using or funding an IUL.

Typically, those in the military will start off with some sort of monthly allotment, as with any savings plan. But that in no way restricts you to paying that same fixed amount forever. Nor are you limited to certain years and certain amounts. You determine the terms and contribution levels depending on how you structure the policy, not the government.

Your Money Grows without Risk of Losses

As I said in the chapter on compound interest, you should only put your money into things that you understand. That's one of the reasons why I am going into the IUL in such detail. I want you to understand how it works. With the IUL, growth of your assets is not based on some Board of Directors declaring and paying a dividend. It's based on the upside of the open equity stock markets. With an IUL, you only execute a market trade if there is a potential profit to be made. If there is no profit, then you don't trade. This means that when the market goes up, you trade and make money. When the market goes down, you don't trade, and therefore you don't lose any money. It's that simple. This is a huge differentiator between an IUL and your traditional retirement plan. In the traditional retirement plan, if the market goes up, so does the value of your account. If the market goes down, so does the value of your account. In the case of a major market correction or crash, this can be catastrophic, especially in those years right before or right after you intend to retire and start withdrawing funds.

What makes this kind of risk-free market growth possible is called a long-term options contract. An option is a contract with an expiration date in which one party agrees to purchase an asset (in this case, a stock or an index) from another party at a specified price and by a specified time (the expiration date). During that specified time, if the stock or index rises in value above the option price, then exercising the option represents a profit for the buyer because they are purchasing the asset at a price that is below market value. In this situation, the buyer would want to exercise the option. On the other hand, if during the specified time the stock or index declines in value below the option price, then the buyer can decline their option to purchase the stock by simply letting the option expire, thus avoiding the market loss incurred by people who already own the stock.

An option is one of a handful of different kinds of financial contracts that are known collectively as derivatives. A derivative is a financial contract between two parties which gets its value or price from an underlying asset like a specific stock or an index fund. There are risky derivatives for sure, like what happened in 2008. There are also people who gamble by selling options on stocks that they don't even own, and who then suffer massive losses when they have to buy a stock at a market price that is much higher than the price of the option and then sell it to someone else for the price specified in the option in order to fulfill the terms of the contract they made. The derivatives used by insurance companies, however, are not gambles. They are very safe, standard financial instruments that have been around for decades, and these insurance companies leverage them in a very conservative way.

Because holding an options contract doesn't require you to have your money actually invested in the market, you aren't continually taking risk on things that you simply aren't going to understand. Instead, you can let the big insurance companies with billions of dollars who are playing a very, very low-risk, long-term game execute those options with your money on your behalf only when they are profitable. Any gains come from the same index that your 401(k) is invested in, but without the risk of having your money being in the index. Using an option to purchase stock is basically having an insurance policy on your purchase.

How Do the Insurance Companies Make Money?

Here's how the money flows. For every dollar that goes into an IUL, around 95 cents is going into the company's general account. A portion of the money in this general account stays liquid so that it is available for participating loans, withdrawals, and the payment of death benefits. The rest of it goes into investments. We are talking about very stable companies, most of them are well over 100 years old, with tens of billions of dollars in assets and A credit ratings across the board. Their longevity is part of the big picture. These companies are playing a very long game. When they receive the premium dollars from policyholders, a portion that they invest goes into an A-rated government and corporate bond portfolio that has a very long duration: 20, 30, 40 years. These are extremely secure bonds. The United States would pretty much have to cease to exist as a country in order for the insurance companies to lose money on these bonds. The longer the duration of these bonds, the higher the interest rate, so these companies are able to keep the hundreds of millions of dollars a year that comes in from policyholders invested in long-term corporate debt at a rate of around 4% to 5%. That's the primary way that these companies make their money.

Remember, the 95 cents is still yours, what I just described is how they are directly investing it for . low-risk, long-term growth. Their goal is to make money for the company, which you, as a policy holder, are a part owner of. So life insurance companies actually like to help their clients. This isn't like what most people think of when they're fighting with an insurance company to pay for healthcare costs or some other damages. Those companies are public and answer to their shareholders for profit. Most life insurance companies just don't work that way.

So what happens to the other nickel on every dollar?

Two or three cents of it goes to pay for the actual cost of your insurance. Exactly how much depends on your health. The better your health, the lower the cost of the insurance. The cost of insurance is actually quite cheap in the long run.

The remaining two or three cents of every dollar goes to pay for these one, two, or three-year options contracts on an index like the S&P 500. The cost of these options contracts is based on the dollar value of the amount that you're purchasing. They work out to something like two or three cents on the dollar. So instead of having the full dollar amount of the value of the contract invested, they only invest and only stand to lose, if they don't exercise the option, a few cents on the dollar. At the end of an option term, if the value of the index on which the company bought the option has gone up, then they execute the option. They buy the index below market value, immediately sell the stock, and then credit your account with a portion of the profits from that sale. I'll talk about exactly how big of a portion in a moment. If the value of the stock(s) on option has gone down, then the company doesn't exercise the option, and they simply let the contract expire, and they are only out the small cost of the contract. Because neither they nor you ever actually owned the stock in the index during the option term, there is no loss. This is how money invested in an IUL can benefit from stock market gains without any risk of market losses. You never actually own any stock, and the insurance company only owns it long enough to resell it and pocket the growth.

In addition to making money from having their assets invested in the super stable, really long-term government and corporate bonds that I talked about above, they also make a little money from exercising these options on your behalf, typically about 10 to 15%. So as an IUL policyholder, you get between 80% and 85% of market gains on these options credited to your account as interest when the market goes up, but you never lose any money when the market goes down.

The beauty of using an IUL for our financial strategy is that your money is in a safe place, where it is continually growing, while at the same time remaining liquid and accessible to you at any time if and when you need it.

How to Access Your Money

Which brings us to the question, how do you access the cash value from your policy? There are two options. You can simply withdraw your money, like you would from a bank, a mutual fund, or any other investment. If you do that, however, then you will miss out on the market growth we just talked about above. In order to access your cash while still allowing that same cash to grow in value, you use something called a participating loan. A participating loan allows you to access the cash value of your account while it is still growing in value, as if it had never left your account.

When people hear the word loan, they immediately think of making payments and paying interest, but you do not pay back a participating loan. Nor do you pay interest on it out of pocket. The interest on a participating loan is typically between 3% and 4%. Right now it is about 3.25%. This charge comes out of the balance of your policy, not out of your pocket as a loan payment. When you pass away, whatever money you've taken out of your account as a loan is paid back to the insurance company out of the death benefit amount. Once the loan(s) is satisfied, whatever is left is paid to your beneficiaries.

This is called a loan because your own money stays invested, and the insurance company loans you the money that you would otherwise have to withdraw from your account. Market returns typically average 7% to 8%, and remember, you can't *lose* anything with your money invested this way, so if you are earning around 6% (80% to 85% of the actual growth), then even with the interest you are paying on the loan, you are still generating more in income than you are spending in interest. In this way, your money can be working for you in two places at the same time. How cool is that?

Here's a real life example of how this works. A policyholder set up his account at age 50. He put a total of $320,000 into his policy over a funding period of 16 years at the rate of $20,000 a year. At age 66, he wanted to start accessing some of the accumulated cash surrender value of his policy to generate some income, so we went into the system to determine how much money this client can take out of his policy indefinitely for the rest of his life without running out of money. At that point, the cash surrender value in his account was $429,000—that's how much money this policyholder can potentially tap into. This amount is the combination of the $320,000 he contributed, plus interest credited to his account as market gains executed on his behalf by the insurance company. In this case, we determined that he would be able to withdraw about $42,000 a year in participating loans without danger of running out of money during his lifetime.

He then began to withdraw $42,000 a year in participating loans as a source of retirement income. When he took out that first $42,000 loan in year 17, the cash surrender value of his policy dropped from $429,000 to $416,000 that year. It only went down by $13,000, *not* by $42,000. Why? Because he took a participating loan from his account instead of a withdrawal, he is still earning interest growth on the entire $429,000. As long as he only takes participating loans and never actually withdraws any of this money, he will continue to see growth on the entire amount no matter how much he takes out in participating loans.

Let's say this individual passes away at age 84. He would have contributed a total of $320,000 to his account during the 16-year funding period. At the rate of $42,000 a year for nineteen years, he would have pulled out a total of around $800,000. That $800,000 in participating loans will be paid back out of his death benefit, and there will be additional death benefit remaining for his beneficiaries.

Access Your Money at Any Time

In the example above, a policyholder started pulling fixed amounts out of his IUL annually as a retirement income stream, but you don't have to do it that way. You can access different amounts of money at any time. You don't have to be 59 ½, and you can use the money for all kinds of different reasons, not just as a retirement income stream.

For example, let's say that this same policyholder decides five years into the funding period that he needs a new car, and he picks out a BMW. He's 54. He's putting $20,000 a year into his IUL, and he's planning to fund it for another 11 years. What are his options for purchasing the car? He certainly has the option to go to a bank to borrow $60,000 and pay them 5% interest to do so. What's another option?

Well, he can access up to about 90% of the cash surrender value in his IUL at any time, for any reason. The reason that you have to leave 10% of the value in the policy is to keep the account open and insurance available. If you take more than 90% of the value out of the account during the first ten years, you would pay surrender charges, and if you cancel the policy and just withdraw all of your money, you risk losing many of the tax advantages of the account. In this scenario, five years into his funding period, his IUL has a cash surrender value of about $85,000. He could borrow the $60,000 from himself and choose to pay himself back instead of making monthly payments to the bank.

I offer this example because I want to reiterate a couple of things. First, it's your money, and you can access it at any time. Second, you don't have to withdraw your money and lose that growth opportunity. You can borrow it, continue to see growth on it, and then pay it back into the account. Third, you have the option to never pay it back at all. However, this example is still early into the funding period. $85,000 is a much smaller number than $429,000, so when the policyholder takes out $60,000, while his cash surrender value won't drop by quite as much as $60,000 that year, it will drop considerably, since $85,000 can't earn as much growth as $429,000. In this scenario, if he wanted to start taking income down the road, that income would have to be much smaller.

The point, though, is that you don't have to just save for retirement. You don't have to sacrifice now for later. You have access to your cash at any time, if and when you need it. What if your house needs a new roof? What if you or someone in your family gets sick? These things happen. The IUL gives you total flexibility to deal with whatever life throws at you.

How the Death Benefit Works

When we first set up an IUL, we typically structure it so as to provide the lowest possible death benefit at the policy's inception. We do this because the smaller the death benefit initially, the lower the cost to fund the plan. In this way, we can bring the typical timeline of a whole life policy, which takes 10 to 15 years just to break even, down to three to five years, if not sooner. This means that it takes a lot less time for the cash value of the policy to grow at a pace that outstrips inflation and quickly starts keeping pace with things like the S&P 500. The money you are using to fund the plan not only keeps pace with inflation, but grows in value over time, because it is invested and is tax deferred.

The amount of death benefit you decide on for your policy will depend on your objectives. If you want to use an IUL as a cash accumulation vehicle, then the best way to make that vehicle as strong as possible is to make the policy as efficient as possible, and one way to do that is to reduce the death benefit, since that's really where the cost is. We really look for a balance between the death benefit protection and that cash accumulation value, and we move those levers up and down as needed over time to manage the policy and mitigate costs. The younger and healthier you are, the less difference lowering the death benefit is going to make, but for someone in their 60s or 70s, reducing the death benefit can drop costs by thousands of dollars. Again, this is an option, not a requirement.

There are a few ways that you can increase the death benefit later as you build this plan. The most cost-effective way is to use both options together in some combination. The first part is to think about the offensive side of this strategy, everything we've been discussing in the last few chapters. What's the least amount of death benefit we need to buy in order to be able to contribute the amount of money you want to use to fund the plan? Then, we increase that death benefit each time you put money into the policy during the funding period. Not to buy more insurance, but simply to keep a fixed window of the amount of death benefit you are buying that is always within the IRS' limited funding amount. Then, when the funding period is over, you have the option to level and/or reduce the death benefit, if you choose. It's an option to reduce costs and focus on the growth. Alternatively, you can continue to allow the death benefit to increase depending on the performance of the underlying index strategy. It's really up to you, and you can decide whenever you want.

The other trick to getting enough insurance on the defensive side of our approach is accomplished with convertible term insurance, the modern version of what should really be called "death insurance." The amount of death benefit you purchase is always going to be the biggest driver of cost. So the cheapest way to pay for a death benefit to hedge against the event of a catastrophe is with low-probability term insurance. That's why it's cheap: there's a low probability they'll ever have to pay your beneficiaries. But thank God it's there when they do. I'm not telling you not to buy term insurance, it's just that we want to be smart about it and use term insurance as part of a bigger picture. So really, while yes, getting an IUL with the amount of death benefit needed for a worst-case scenario in order to protect your family is good, what's even better is locking in the insurability and health rating of that individual for a certain amount of death benefit at a much cheaper cost. This is huge and often overlooked. Most people think they'll buy more insurance later in life when they "actually need it." It's like, dude, you need it now because it'll give you access at low cost. Later in life is when people find out they're uninsurable because of something they said to a VA psychologist once. It's nuts.

Locking in the death benefit amount is a huge advantage because now you have that much more you can purchase in future IUL contributions when you want to fund it more, but your first policy is already at the IRS tax limit. Now you can just transfer a portion of the term death benefit by converting it to an IUL, all without having to apply for insurance and go through underwriting again. And of course, now you'll also have more death benefit in a permanent policy meant to leave a legacy.

There's never been more flexibility in insurance before. And there's definitely never been a better insurance strategy to give you so many options for a return on your investment without stock market gambling, without high investment costs, and most importantly, without having to pay Uncle Sam a dime in taxes. You already served for it. Why pay double?

The best part is that you can get a tax-free return on your investment, without having to die!

A Tax-Free Income

Another important aspect of the participating loan feature is that you do not pay taxes on money that you loan to yourself. Why? Because you don't pay taxes on loans. Think about it. When you get a loan for a car, do you have to pay taxes on that money? Do you have to pay taxes on money you borrow as a mortgage? Of course not, because you are ultimately going to pay that money back. Since you're not keeping it, it's not technically income. The same thing happens with a participating loan in an IUL. The money will eventually be paid back out of the death benefit on your policy. Whatever is left over of that death benefit will then pass to your beneficiary, again, tax-free, because the death benefit of a life insurance policy is not taxable either.

According to the IRS tax code 7702[1], insurance products can grow not only tax-deferred, but you can access both your money and associated growth without paying any more taxes. This means that assuming the policy was properly funded within the generous IRS-mandated guidelines, you can access the money in an insurance policy tax-free. This is a hidden secret. It's really cool and easy to set up, but it must be funded properly.

As you can see from the articles on my website and from my YouTube videos, I have done the research and developed a team to make something unique that solves a problem when it's designed properly. I can't vouch for or endorse a product or strategy done by any other financial advisor. They may use it their own way, for a variety of reasons, and I use it for the reasons I outline in this

1 https://www.investopedia.com/terms/s/section-7702

book as the foundation for financial control. So, if you work with me, then we teach you how to fund it properly. Whether you do it or not is always up to you.

Additional Benefits

Pay for Long-term Care

People are living longer. They are acquiring additional health costs that didn't exist just a few decades ago, but not all of these costs will be covered by the VA and Medicare. Among those aged 65 and over, 69% will develop disabilities before they die, and 35% will eventually enter a nursing home. What if you need assisted living? There's no option to leverage the SBP death benefit to help with long-term care costs. Neither do term life insurance or whole life offer any feasible long-term care option.

An IUL offers something called an accelerated benefits rider. That means that you can access the death benefit on your policy while you're still alive to pay for your health care costs. There is no cost for this rider. The difference between an accelerated benefits rider and taking participating loans is that the qualifying payments can come from the value of the death benefit rather than from the surrender value of the policy. This rider is a way to use the death benefit early without using a participating loan. So if you get sick and you need a nurse to come to the house, or if you go into a nursing home or whatever you need to do, you can tap into about 2% of the death benefit on your policy tax-free and penalty-free on a monthly basis to help pay for those costs. For example, if you have a $500,000 death benefit on your policy, 2% of $500,000 is $10,000 a month. That's pretty good money, and again, it's tax-free. You can even use this rider to pay assisted living or long-term care costs for a close friend or family member. As with the participating loans, whatever is left over of the death benefit after you pass away is paid to your heirs.

Litigation Risk Protection

A permanent life insurance policy is not considered part of an estate tax bill or otherwise considered to be part of an individual's net worth in terms of what lawyers are able to retain or what can be tapped to pay for legal damages. In the event that you are involved in a lawsuit that ends badly, whatever other assets might be taken from you, you will retain complete control over and access to your IUL.

Is It Too Good to Be True?

While it may seem like the advantages of an IUL are all on your side, these policies are very, very profitable for the companies who issue them.

First, the fact that an IUL is life insurance means that they are going to underwrite you, and the underwriters are very smart about estimating how long you are going to live. They're going to talk to your doctors and look at your medical records and your prescriptions, and from that they are going to extrapolate longevity.

Second, these insurance companies only have about a 10% attrition rate. What does that mean? It means that 90% of the people that do business with them hang on to the policy for the rest of their lives. So these companies know that if they do a policy with you, nine times out of ten you're not going to go anywhere. This is very different from what usually happens in our industry, where people work with an advisor for a few years and then move their money to another company and then another. People do the same with bank accounts. But these insurance companies are far more stable than banks. This particular company I was looking at in the example earlier in this chapter has been around for 135 years and has $75 billion in assets just in their general account, and they are A-rated across the board. They are very, very profitable, very old-line, very conservative companies, so that's part of that equation.

Based on all those factors, the insurance companies have a really good idea of how long you're going to live, and they know that you've got a very high likelihood of being comfortable and satisfied with their product and will leave your money there. The fact that they are able to take the big pool of assets that they have and invest in the very long term is how they can make these attractive returns available to us.

Who Can Most Benefit From an IUL?

The connection between the IUL and the solution for the military service-member and veteran is not that it's an across-the-board better solution, or one-size-fits-all, or should be used by every single person. We've already alluded to the fact that on average, these solutions are for those aiming to be in the top one percent. It's not for everybody. It's not for those who just want the absolute basic necessities, and who don't want to think about how to do things like life insurance coverage while they're in the military. That's why SGLI is easy, relatively cheap, and standardized for everyone. But IUL might be a great tool to complement something like an SGLI for those who:

- Don't want to just to rely on the government
- Don't want to just rent insurance for as long as they need it until they can deal with it later
- Don't want to be limited by the amount of coverage they have
- *Do* want to save a little bit more money in a tax-advantaged vehicle, with no risk, so they will have money available to leverage at their transition (whenever that may be) to use for whatever reason they want

For the kind of person who is very comfortable with a safe and steady income for life, and who hopes that retirement will work the way it has for the generations before them, we have all been sold the story that taking advantage of these tax-deferred retirement accounts, including TSPs, 401(k)s, and Roth IRAs are the way to do that. But I just don't have faith in recommending this course of action, like the Dave Ramseys and typical financial advisors do, because nobody can say with any certainty how that is going to work out over 30 or 40 years. To each their own. But for those who have other serious goals beyond just building a nest-egg for retirement, for those who want to be financially stable and secure, who want to maximize their opportunities at the end of a military career, but who intend to pursue a lifetime of service, then having only retirement accounts isn't the best way to accomplish that goal, is it? This is where an IUL—with tax-advantaged growth, no rules on how to access your money, no strict restrictions on how to fund it, and no risk if there is a stock market crash—can be a great place to put money to prepare for any opportunity that presents itself.

Taking advantage of this product requires someone who can both fund it properly and understand how to structure it properly. Most financial professionals haven't learned how to do either because they make so much money using the old tools, they didn't bother to keep up with innovations in the industry. They haven't learned these new tools. Not only that, they can't charge fees off them like wealth managers do, especially the ones who manage all these retirement assets, and that's how they make their money. Why would they agree to recommend something that doesn't pay them over the long term and charges clients less over the long term?

I am a licensed life insurance and investment professional. By definition, as an Investment Advisory Representative (Series 65), I am a "fiduciary." But my definition of fiduciary goes far beyond the standard advisor responsibilities for asset allocation and risk tolerance. If the definition of a fiduciary is to "do what is in your client's best interest," then for those who work with me and who are serious about financial control, there is no perfect asset allocation that will be in their best interest because they give up far more control than they realize when they invest in traditional retirement accounts like the TSP, IRAs and 401(k)s.

Sometimes, I wish I had learned another way to make a living when I got out of the military, but traditional financial planning is what I learned, and as I have explained in this book, I hated it. I found the way that most financial professionals conduct business to be just plain wrong. I could have just quit and tried to figure out another way to earn a living, but I didn't. Instead, I went on a journey to figure out a better way to serve my clients and provide value. That's what this book is all about. That's what US VetWealth is all about. If you want to try a different approach to post-military life, my approach still offers all the same benefits as the status quo approach, but without the status quo drawbacks. It still allows you to create retirement income for yourself. It still allows you to protect your military pension for your family should something happen to you, which is highly unlikely. It allows you to benefit from stock market growth without putting your assets at risk, and oh, by the way, it also allows you to never be taxed, ever again.

Final Words About Asset Protection

It is now more accessible than ever for the middle class to participate in what the wealthy are doing. I don't mean using the IUL specifically, at least not yet. Most wealthy aren't even aware of this, but they soon will be. I'm talking about how you can use this to better position yourself to do things that the wealthy do to create more time, income, and legacy for themselves and their families. They do this through constant self-development, identity creation, and financial control. Why are so few of the 1% who served doing the things that the wealthy 1% are doing? There are no other employed individuals with the opportunities or assets that veterans have, which means that veterans can very much set themselves up to join the wealthy 1%. Veterans deserve more than to be talked down to like they are disadvantaged and under-informed. It's not that hard for them to understand their finances, and in the next chapter, I'm going to walk you through a case study to see how using an IUL can provide value and open up more opportunities for service-members and post-military life.

20

A US VetWealth Success Story

Creativity takes courage.
~ Henri Matisse

Before I leave you with my final thoughts, I am going to show you how a real client has used my Wealth and Liberty Strategy to change how he views himself and his future, improve his financial situation, and expand the opportunities he has available to him post-military in the years he still has left before complete retirement.

Commander Jones' Story

Commander Jones is a 40-year-old Navy Officer, a supply logistics guy. Not too long ago, he was assigned as a unit transition assistance officer in the U.S. Navy. He found this quite ironic, considering he was far from transition himself, but such is the military logic we've all come to accept.

For anyone who might not be aware, each command has a transition assistance officer that helps mid-grade officers who are transitioning out with no benefits post active-duty, as well as those who are retiring out of the service with full benefits off of active duty. Commander Jones worked in this position for three years, During that time, among the many junior sailors he helped, he saw four officers transition off active duty who were what's called "failed to select," which means that they didn't promote "on time" to the next rank. He also helped five other senior officers, O5s and O6s, retire out of that command while he was there.

What he saw that all of these nine officers had in common was that none of them were ready for post-military life. Commander Jones' job was to shuffle them through the bureaucratic, check-the-box steps of transitioning out of the military. Contrary to what you may be thinking, it was not this experience that woke Commander Jones up to the fact that his own transition, too, was inevitable. That didn't happen until the day that he received his notification that he had been selected for promotion to commander, O5. That's when the reality struck him that everybody gets out of the service someday, and he thought to himself, "Why am I not more ready?" He began to think about the other officers he had helped through transition and thought, what if he himself hadn't been promoted from O4? What if, in two years' time, the military decided not to keep him, and he, too, was thrust into transition? That was the day that he realized that just trying to check the boxes and promote to the next rank was not the right decision for him and his family. He had what I called in my first book, *Life, Liberty, and Purpose,* his "Ah-Ha" moment, and he began to think about when and how he would ultimately separate from the military.

The first thing he thought about on that day was his financial situation. This drove him to take a 30-day challenge about getting his finances in order. This challenge had been created from a military perspective by someone who had previously retired from the military. Several months after that, in this mindset, Commander Jones met me.

That was a year and a half ago. At that time, he was assuming that he was going to retire from the Navy in a few years, get his pension, and then go work for a big logistics company. He was already fairly active on LinkedIn and fairly well-networked, and he accepted one of my offers for a free LinkedIn profile audit. So I did a 15-minute screen recording while I went through his profile and pointed out some opportunities that he was missing and some different ways to think about how he was doing things and why. I explained to him how if he improved the way that he uses LinkedIn today to build his personal brand and really start networking, that he could completely change his prospects for where he might be in the next three years.

Commander Jones has done all the "right things." He's been saving for many years. His wife is a stay-at-home mother, and they have four kids. They're living primarily off of just his salary, so he was concerned about making sure he was prepared for what would come after the military. His concern prompted him to ask me, from my perspective as a veteran who has already been through the transition process, what I know about military transition-related benefits as they relate to personal financial planning. Specifically, he wanted to know, "What is it that the military isn't telling us?"

Well, as I have already said in this book and written about extensively on my website, when it comes to our retirement briefings, there is a *lot* that the military isn't telling us, especially for a guy like Commander Jones, who hasn't even had his retirement briefing yet. Some of these things maybe the military should be telling us, but not everything. After all, why is it the military's responsibility to train you how to get a job? Corporations in the civilian world don't feel any responsibility for ensuring that when people leave their employment they are able to succeed in their next position. When you look at it from that perspective, the military really is going above and beyond what's probably necessary just in doing what they do. But even more than that, why should we expect the military to be good at keeping up with the way that our economy and the opportunities available within it are changing, when things are changing so rapidly? It's not the military's job to educate us about how to get on with our lives, and we can only hold the military responsible for our lives post-military up to a point.

So we started talking about the SBP, VGLI, and the TSP, and how all of these things are super-limiting, one-size-fits-all solutions to financial planning that funnel veterans straight into mediocrity. We talked about how veterans come out of the military with the mindset that they have to get a job, how in Commander Jones' case, it was totally predictable that a Logistics Officer would assume that he would get a logistics job post-military and replace the military structure with a civilian corporate structure. I have said already, and I will repeat it again here, that's great, if that's what you want to do. You can get a good paycheck and have a good life. I asked Commander Jones if that was what he wanted to do, and his answer was, "No, not at all. I've been following orders for so long. Maybe I would like to do my own thing now, but I don't know what that is yet." Through the course of this and subsequent conversations, Commander Jones became a client.

Commander Jones' Finances: The Starting Point

The first thing that we did when looking at Commander Jones' finances was to determine the actual amount of his military income. We used the regular military compensation calculator, which is specifically designed to show civilian equivalent pay. Commander Jones thought that his annual pay was $108,270 a year. However, like many on active duty, he hadn't calculated in his tax-free housing and other allowances, which brings his actual compensation up to around $160,000 a year. He's married with four kids, though, so his taxable income is only $83,000, which means that he currently only pays a few thousand dollars in taxes. So while his marginal tax-bracket was 22%, was he paying 22% of $160,000? Nope, of course not. In fact, his taxes are only calculated from the $83,000 and his effective tax-rate, what he actually paid, is less than 1% of his real income if you account for the full $160,000 that makes up his family's lifestyle. If he were to continue along his current path, it would be a huge wake-up call for him when he gets his next job if he had to start paying taxes on the actual amount of his income. That tax will would be closer to $22,000 per year if he made $160,000 as a civilian.

Commander Jones and his wife have been maxing out their Roth IRA at $12,000 a year. In total, they have $150,000 invested in a target date retirement fund. The expense ratio isn't that bad, .60%. He is also maxing out a Roth TSP at $19,000 a year. So he has $31,000 a year going into retirement accounts. Those accounts already have a combined balance of $250,000, invested fairly aggressively (80/20 portfolio) since he still has such a long time until retirement, though frankly, based on common CFP wisdom, he could go much, much more aggressive with that timeline. But let's talk a little bit about what that actually means.

Realistically, to go after that 8% average rate of return, the Modern Portfolio Theory (MPT)[1] states that with an asset allocation of roughly 80% stocks, 20% bonds, in order to get that 8% return over a long period of time, then over any six-month period of time you have to be willing to risk 15% in losses for the potential of a 23% gain. In Commander Jones' case, with $250,000 invested, in any six-month period of time, he would have to risk losing approximately $40,000 for the potential of gaining approximately $60,000. My question for Commander Jones was, is that really okay? Do you really feel comfortable with that? What are you going to do if things don't go your way and your account value drops by that $40,000 in a short period of time?

People always say that they can handle the risk, but the DALBAR study that comes out every year shows that when the stock market starts to go down, people can't handle the risk, and they tend to pull their money out. These kinds of losses aren't that big of a deal if you only have a few thousand dollars in your account, but if we're talking about real money, half a million, a million dollars, people panic and pull their money out of the

1 https://www.investopedia.com/terms/m/modernportfoliotheory.asp

market before they lose any more. The problem with that is that when the market does rebound, the people who pulled their money out of the market and stopped adding more to it don't get the benefit of its upside when the market starts rebounding to the positive again.

A lot of the fear people are having around the market today has to do with it being overvalued, along with all the uncertainty going on. The Federal Reserve is just printing money and putting it into the stock market, so the market keeps going up. There's also fear of missing out. The MPT doesn't say anything other than, "Well, this is what stock market returns have looked like over a long term."

Commander Jones also has $75,000 in a money market fund earning almost no interest because interest rates are so low, as the Fed is pulling another rabbit out of the hat to stimulate the economy. He thinks he needs to have it somewhere else where it can be available in case of emergency, but he also doesn't want to have it earning almost zero growth. He's frustrated. He's afraid to invest it right now because the market is so high, and he wants to do something with it because right now he's not getting any growth on it.

So that's the financial position in which Commander Jones came to me. He had three years left in the military before his transition. Since he has four young kids, he plans to get a job and keep a salary until the kids are out of the house. But he's also planning on starting a business in his free time. He understands that it takes many years for businesses to get going, especially if you're only doing it part-time while still working as a salaried employee for someone else or for the government. Commander Jones is highly employable given his duty career field with the military, and he can expect to get a job with at least $140,000 in salary, income which would be in addition to a $60,000 a year military pension. Unfortunately, that means that he will then have an annual taxable income of $200,000 (as opposed to the

$83,000 taxable annual income he has now), which means he will no longer be able to contribute to his Roth IRA.

Financial Aspects of the Wealth and Liberty Strategy

Though Commander Jones will no longer be able to contribute to his Roth IRA and will miss out on additional tax-free income, the money in those accounts is tax-free savings. He was still planning on funding his 401(k) and maxing that out at $19,000 a year.

His original plan was to get a job and work until he was 60 or 65. But now that he's implementing the Wealth and Liberty Strategy, Commander Jones no longer wants to be anyone's employee after age 50, whether his business is producing income at that point or not.

Let's look at some costs on his investments if Commander Jones didn't make any changes and followed the traditional plan of getting out of the military and then working for another 20 years.

- Roth IRA: Commander Jones is in the target date USAA mutual fund with a current balance of $150,000. He will only be able to fund it for three more years, because after he gets out of the military, his income will be above the Roth eligibility threshold. With all that in mind, 20 years from now, with a much more realistic and conservative growth rate of 6%, he can expect that account to have a balance of around $518,000. His total cost on that account will be about $38,000.
- TSP: Commander Jones was planning to roll over his TSP into a new company 401(k) or something similar and continue to contribute. After 20 years, he had expected to have gotten an average rate of return. For the purposes of estimating the total costs of his traditional retirement account, we estimated a conservative 6% growth, which would have resulted in another $106,000 in costs because the 401(k) fees are so much higher. So when he came to

me, Commander Jones was basically planning to commit to $144,000 of costs for those two investment accounts.

- SBP/VGLI: Commander Jones was also assuming that he would take the standard government benefits when he retired and follow the system he saw his senior peers going through as a transition assistance officer, in which case he would have committed himself to about $150,000 in costs over 30 years to the SBP and around another $100,000 for VGLI if he were to keep paying for it for 30 years. At least with VGLI he has the option to cancel in the future; with SBP, he doesn't.

What we decided to do was instead of putting any new money into the Roth IRAs or the TSP retirement account, we would just leave those balances intact where they were and start putting new money into lower-cost investments. He's going to take that $31,000 a year he was already putting into the Roths and the TSP and fund an IUL instead. For the first three years, he's going to fund the IUL with $31,000 a year from new income while on active-duty. He's also going to use that $75K he had in the money market and split it up over three years. At year three, when he is ready to transition from the military, he and his wife will have $165,000 available to them, tax-free. This is money that could be specifically used for transition, whatever or whenever that may be. Maybe it's 10 years after he separates from the military, when he has a new business idea, and by that time the account is worth $300,000 or more.

We also stopped letting the government withhold his taxes from his paycheck beyond what is necessary to avoid penalties for under-withholding. Instead, we put that money that would have been going prematurely to the government into the IUL. When it's time to pay his taxes, he will take the money out and pay them. But even though he has taken out the money he needs to pay his taxes, as I explained in the last chapter, he is still getting growth on that money in his IUL. In order to maintain the tax benefits, he needed to purchase a $1.5 million death benefit. He also got a $1 million convertible term to help cover the full value of his pension, and give him access to convert more death benefit to the IUL if he needed more room to add additional funding if his business does well.

Once he's out of the military, when he gets his next job, again, he will now use his military pension to fund the IUL, along with what he would have been contributing to a 401(k) for just 7 more years. Commander Jones wants to supercharge funding for his retirement so he can retire early and be done with working for anyone else regardless of how his business works out. At age 50, he's done funding his account. He doesn't put one new dollar into it. At that time, he's got an account worth $871,000 earning interest for him based on the (upside only) performance of the stock market, with no risk of losses. Of that amount, he can access $635,000 at any time. Beginning at age 50, without ever contributing any more money to his account, he will be able to withdraw $62,000 a year tax-free for the rest of his life, essentially doubling his pension. His children will be grown. If his business works, great. If not, he still won't have to work.

Now let's compare the costs. If he had continued down his original path to fund his Roth IRA for 3 more years and his 401(k) for 20 more years, Commander Jones would have paid $144,000 in costs for those products just at the 20-year mark. With the IUL, at the 20-year mark, the total cost is $110,000. But remember what I told you, the IUL gets interest credits after you finish funding the policy, basically reimbursing you for what you paid in premiums. At the 20-year mark, Commander Jones will have already seen $60,000 in interest credit, which basically drops his total cost down to $50,000, so it ends up being much cheaper than the his mutual funds in the long run. Additionally, there are almost no new costs. The longer he has the account, even as he withdraws from it, the investment just continues to grow and the costs essentially disappear.

In His Own Words

After coming to understand the real risks inherent in investing in the stock market, as well as the real risk of future taxes, Commander Jones compared riding these risks to being swept along by the current of a river. He said, "I need to have control. I need to be able to choose what I'm going to do when I get out of the military. I want to be able to choose where my money comes from, where it goes, and how I can use it. I don't want to have to obey all these rules, or have a boss who chooses if I get that paycheck." For those reasons, Commander Jones chose not just to make changes in his financial planning, but to start laying the groundwork for complete autonomy in his post-military life. In his words:

"We're all in transition while we're on active duty. As soon as people leave the basic training pipeline, whether they recognize it or not, they are on a transition mission, and that mission will turn into a battle before they know it. If they're not ready and willing and able and positioned well enough to win that battle, they're going to get their *** kicked. So my goal for my remaining time on active duty and afterwards is to change the paradigm that transition happens within 18 to 24 months of getting out of the service. That's just not true. That's a wrong paradigm. It's a challenge we have today, which I believe can be fixed if we get people on LinkedIn and get them to use it as an efficient tool for networking for their future."

Finding His Purpose

Commander Jones set out on a personal journey to discover his purpose and to find those whom he is meant to serve. He's been on this journey for the last few years or so, and he's spent over a year testing and building his brand. Now, this has been fun to watch, because it's still in the very early stages. He's been trying out lots of ideas, but he keeps coming back to something that's needed in the marketplace because nobody else is really doing it. It's unique and creative. It builds on his own story and his own experiences and knowledge and doesn't require any sort of qualifications or degrees or whatever to come up with this way to add value to other people.

Because he shifted his mindset away from thinking that he just needed to get a paycheck and put money away for retirement, a portion of Commander Jones' assets are now in an IUL, which allows him to access those assets if and when he needs them to launch his business. He has now geared his networking strategy around his idea and is already finding other people doing similar and complimentary but uniquely different things and connecting with them so that they can help each other rather than competing. He has entered the online world where anybody can create a side hustle, and that's all he's doing. He's not even trying to monetize this right now, because he knows, hey, while I'm still in the Navy, I can put a bunch of money away into this advanced equity insurance strategy so that three or five years from now, if I need it for my business or to supplement my pension for a little bit, it will be there.

To Sum Up

We need money. With the Wealth and Liberty Strategy, you get to decide how much money you need and what you're going to do with it. What you don't get to decide is the amount of time that you have left in this world. Everyone has to determine for his or herself whether or not the time-for-money tradeoff that's inherent in traditional financial planning is worth it. How many hours do you spend every week right now for the portion of your salary that's going into a traditional retirement plan? When you get into your 60s, or 70s, or 80s, and you can finally convert some of your money back into time, will the choices that you are making now have been worth it? Will you still be able to do the things that you really want to do? Or will some opportunities be lost to you forever by then? And that's assuming that you are even able to retire. What ends up happening more and more often these days is that people have to keep working.

But guys like Commander Jones are taking control. He's been busting his butt. He learned how to create websites. He learned how to create podcasts and videos. He's learning the art of copywriting. He's building a personal brand and learning how to add value, and coming up with ideas, and networking while he's still in the Navy. I am actually shocked at how much he has already accomplished and how quickly, because when I was in his position I didn't even know this world existed. I bumbled through it for many, many years before I started to see what's really happening.

The fact that Commander Jones recognized opportunity and acted on it while he's still on active duty, with three years to go before he gets out, is going to set him up to be lightyears ahead of most active duty, who don't start thinking about what they are going to do next until they're sitting in their retirement briefing. The steps he is taking now are going to enable him to avoid having to pay 6.5% of his pension for the SBP. He won't have to pay for VGLI, which is not only more expensive than SGLI, but which would cost him over a thousand dollars a month by the time he's in his seventies with a maximum benefit of only $400,000. This is all money that he will now be able to keep, and that he can now use, not only towards something that he wants to do with his life, or for simple living expenses, or to build more equity in his IUL.

The timing here is important. Commander Jones now has these options because he got qualified while he is still young and healthy.

Conclusion

Forty hour workweeks are a relic of the Industrial Age. Knowledge workers function like athletes—train and sprint, then rest and reassess.
~ Naval Ravikant

This book is something you have probably never encountered before: a financial book that is *not* only about money. It's about how to set yourself up for your future rather than just trusting that Wall Street and the government will be there for you 30 years from now. It's about identity and transformation. It's about getting to a clear-eyed understanding of who you have been, what you want, and who you want to become. It's about moving from budgets and regulations to financial independence; from grinding to thriving; from paychecks and benefits to wealth and liberty. And like all journeys, it begins with a single step.

Why is it that people turn to technology for everything, yet in the financial world we're still supposed to keep doing things like it's 1984? Right now we are living in what Klaus Schwab, the Founder and Executive Chairman of the World Economic Forum, dubbed the Fourth Industrial Revolution. That's important. The fact that enough has changed, and changed enough, that the times that we are living in require a new designation is important; but equally important is the fact that we are in the midst of this Fourth Industrial Revolution right now. It is still unfolding, which means we don't know how it's going to end up yet or where it's going to lead us.

I like to look back at history, because understanding where we have been helps us to understand how we got to where we are. Sometimes things that are going on in the present moment don't always make sense in the present moment without the perspective of history. The post-military retirement myth that I talked about in the Introduction to this book is one of those things. Consider:

The First Industrial Revolution began towards the end of the 18th century and ushered in the 19th. It was about mechanization. It revolutionized agriculture, and the invention of the steam engine brought us the railroads, speeding up the economy. The Second Industrial Revolution happened about a hundred years later. This revolution was largely about energy (electricity, gas, and oil), and it brought rapid changes in transportation, construction (steel), and communications. The Third Industrial Revolution began less than a hundred years after that, in the late 1960s. This is the revolution that brought us nuclear energy, electronics, telecommunications, and computers. We began to explore space. The use of automation expanded. All of this paved the way for the Fourth Industrial Revolution, where we are today.

As you can see, the leap from the Third to the Fourth Industrial Revolution only took a handful of decades. This is important because it shows us that the more technological change that we experience, the more rapidly the next set of changes is likely to arrive. But it's also important because this most recent Industrial Revolution has come up on us so quickly that some of the people witnessing it are the same people who witnessed and participated in the third one. What does this mean? It means that while large parts of our society are changing, other large parts, which are still largely occupied by the people of the Third Industrial Revolution, are *not* changing. Not every industry or organization is necessarily keeping up with the times.

I started this book by talking about the post-military retirement myth: Translate your military experience into the skillset language that the civilian world will recognize. Write your resume. Put on a suit and go to job fairs. Get hired by a government contractor and spend the next 20 or 25 years of your life following their orders, all the while socking away as much money as you can into a retirement fund that you can't touch until you're in your sixties. If you do all of these things in exactly this way, you won't be doing the "smart" thing, you'll be doing the smart thing. You will be successful, and you will be safe.

This myth belongs to the Third Industrial Revolution, the one that dawned at the end of the 60s, the baby boomer generation. It continues to be so pervasive today because there are so many aspects of society where the baby boomers are still either in charge or have enormous influence. When you look at the situation in the historical context of these successive industrial revolutions, it's much easier to understand that a lot of organizations and a lot of people are still living in the past.

When the baby boomers were entering the workforce, you could still learn a skill and then do it for the rest of your life. You could get a job with a company back then and work there for 20 or 30 years and then retire with a pension. It was possible to achieve a kind of stability then that, quite frankly, is just not possible for anyone making those same status quo choices today. Yet, the guidance that our veterans receive as they prepare to exit the military has not kept up with the times. The DoD is preparing our veterans to enter an economy that doesn't even exist anymore.

Most of the baby boomers are not our leaders, no matter what elevated positions they may still hold. Those who are still doing things the same way that they did them 20, 30, and 40 years ago represent the lagging aspects of our economy, the sectors or portions of sectors that have not yet innovated, as I have shown you is the case in the financial and insurance industries.

I wrote this book for active duty military and veterans who want to be part of this Fourth Industrial Revolution and for those who are likely to bring about the fifth. It's no longer enough to be a dependable cog in some company's economic wheel. The dividing line between employees and entrepreneurs is beginning to blur and fade. Companies are no longer looking for mechanics, monitors, and managers. They are looking for innovators, problem-solvers, creators, and leaders. The companies most poised to succeed over the next five years are already shifting their focus when it comes to how they hire and train employees. They are requiring less and less specific qualifications, degrees, and years of experience in favor of creativity. They are also increasingly outsourcing tasks and projects to independent contractors and/or freelancers operating virtually. Whether you want to be an employee or an entrepreneur, you will have to have the same kinds of skills in order to succeed.

If you are feeling doubt, fear, or confusion about what you are doing or going to do post-military, take that as a sign that you are tuned into the fact that the world has changed. These feelings aren't an indication that the things your transition advisors are telling you are too complex for you to understand, they're a sign that on some level you know that they don't make sense.

In this book, I have covered a lot of territory. I have asked you to reconsider everything that you have ever thought or believed about the world. I've asked you to look inside yourself and to recognize or rediscover your true purpose. I've asked you to take a leap – but not of faith. The information that this book presents about the state of the economy and what those who put all their resources into the status quo can expect to happen in the coming decades are not fringe speculations. They are facts. The economy has changed. The old rules no longer apply. The big winners today are not those who are playing by the rules of the establishment, but those who are finding those whom only they are meant to serve, offering them value, and then creating and maintaining their own little economic ecosystems. This applies to you whether you want to be an employee, an entrepreneur, or an independent contractor.

Everything that I have written in this book comes back to a single concept: personal power. Recognizing and exercising your own personal power is an essential component of happiness. The future belongs to those who can do this. And as a veteran of the U.S. military, you have more options and are better poised to find your unique place in the modern economy than anyone else in the country.

In the Introduction, I asked that you spend a few minutes to think about the following:

- What dream have you been putting on hold while you served your country?
- What value do you now want to add to the world?
- What passions can money help you to enjoy?
- What do you want your life to look like beyond the military?

I hope that by this point you have given a lot of thought to these questions. I hope that you have started to have some ideas and that you have thought of a few things you might like to try out. Maybe you have even taken some steps in a new direction. If and when you do, you will be setting off on the most consequential journey you will ever make.

I didn't like making money in the financial industry the way I was told to do it. That's because your finances aren't about money. They're about your life. I wrote this book because my approach is too important to let our veterans risk missing it. I had to stand up and share it, so I hope it was helpful. Share it with a friend and help me spread the word. Let's keep talking about it until no veteran ever has to get to the end of their life filled with regret that they never got their identity back.

About the Author

Scott R. Tucker is an author, speaker, and the founder of US VetWealth, a lifestyle and financial consulting brand that helps service members go from paychecks and government benefits to wealth and liberty. He likes to say, "I help the 1% who serve our country become the 1% who influence it." A West Point graduate, serial world traveler, military financial expert, and entrepreneur, Scott brings valuable experience and insight to those who have sacrificed so much in service to our country.

Made in United States
North Haven, CT
31 January 2026

87716104R00188